Camouflaged

Camouflaged

Forgotten Stories from Battlefields

Probal DasGupta

JUGGERNAUT BOOKS
C-I-128, First Floor, Sangam Vihar, Near Holi Chowk,
New Delhi 110080, India

First published by Juggernaut Books 2023

Copyright © Probal DasGupta 2023

10 9 8 7 6 5 4 3 2 1

P-ISBN: 9789353453459
E-ISBN: 9789353456580

The international boundaries on the maps are neither purported to be correct nor authentic by Survey of India directives.

The views and opinions expressed in this book are the author's own. The facts contained herein were reported to be true as on the date of publication by the author to the publishers of the book, and the publishers are not in any way liable for their accuracy or veracity.

Chapters 7 and 9 – While the events in these stories are based on true incidents, all the names of individuals mentioned in these chapters have been changed to protect the identities of the real people involved.

This book contains some reconstructed or modifed dialogue based on available historical evidence and sources.

All rights reserved. No part of this publication may be reproduced, transmitted, or stored in a retrieval system in any form or by any means without the written permission of the publisher.

Typeset in Adobe Caslon Pro by R. Ajith Kumar, Noida
Maps created by Mohammad Hassan

Printed at Replika Press Pvt. Ltd.

Nisha, my wife and the spouses of those who serve in the line of duty.

Contents

Introduction	ix

Part I: Indian Soldiers, Foreign Wars

1. Sultans of the Sky — 3
   Hardit Singh Malik and Indra Lal 'Laddie' Roy

2. Message in a Battle — 34
   Gobind Singh, VC – the Hero of Cambrai

3. Three Lives in a War — 58
   The Miraculous Escapes of Chanan Singh Dhillon

Part II: Defending the Borders

4. The Boy Who Would Become Stak — 93
   The Legend of Chhewang Rinchen

5. Rise after the Fall of 1962 — 119
   The Amazing Comeback of Haripal Kaushik

6. Top Guns of Boyra — 147
   Gentlemen at War

Part III: Modern Era: Unseen Adversaries, Identity Wars

7. A Bloodless Pact to Victory 179
   An Unlikely Alliance

8. Warrior's Code of Courage 210
   Leading with Heart and Honour

9. The Militant and the Major 235
   Finding Nizamu

10. Nariman House, 26/11: A War Comes Home 255
    Of Bravehearts and Their Families

Notes 290
Select Bibliography 309
Acknowledgements 313
A Note on the Author 318

Introduction

On a wall of the Pithoragarh fort in Uttarakhand, there is a marble plaque that tells us that 1,005 men from one village went to fight in World War I (WWI) in Europe in 1914. Later in the decade, Honorary Lieutenant Rudra Bir Sen from Pithoragarh would take part in the Third Anglo-Afghan War in Waziristan in 1919 and receive the Indian Order of Merit for bravery. Around 90,000 Indian soldiers died in these wars. Ninety years later, Rudra Bir's grandson Lieutenant Colonel Sundeep Sen led the action against terrorists at Nariman House during the Mumbai attacks of 2008. Incredible stories from battlefields both far away and at home have spanned many generations of soldiering in the Indian Army.

In fact, there are many Indian families where several generations have served in the army for over a hundred years. Brigadier Saurabh Singh Shekhawat's great-grandfather fought in WWI. Stories of his battles in the war were passed on to Saurabh by his grandfather. An inspired Saurabh would continue the tradition of serving in combat units.

As a kid, Colonel Rajendra Singh Rathore would look agape at a portrait of his grandfather Gobind with medals emblazoned on his chest. Rajendra's father was five years old when Gobind died in 1942, but Rajendra grew up hearing about his grandfather's incredible bravery in the Battle of

Cambrai. Rajendra, his father and grandfather Gobind served in the 2nd Lancers regiment – ensuring a generational continuity in the same regiment over a century.

It isn't uncommon to find oneself listening to fascinating tales in messes and regimental gatherings in military cantonments. Many of these stories are not publicly known. Others are in danger of being forgotten altogether. These stories range from the renowned stands to eccentric little tales. However, most of these stories have remained neglected, and occasionally been recorded in footnotes, featured in articles, family tales or regimental histories. They are seldom mentioned in history books. This book is a modest attempt to fix this.

The Indian Army, among the most versatile, busiest and inclusive of all armies in the world, has had an enviable history of participating in several major wars in the last century beginning from WWI to the more recent ones on terrorism. Soldiers from different regions, religions and economic classes across India have been a part of the now nearly 1.5 million military. A glorious range of stories have spawned over a hundred years involving incredible experiences of Indian soldiers in battlefields across continents.

This book comprises ten unique stories in three parts, beginning from 1912 to 2008. These stories from different periods mirror the journey and changing priorities of the nation over a century. They reflect the various moods and circumstances of the eras they are nestled in.

My book doesn't claim to cover all the forgotten stories. Among the stories in the book, a few have been mentioned briefly in regimental histories and newspaper articles. Some of the stories haven't been published at all. While some of the stories have featured in a different form in earlier books, they have been retold here distinctively, using a different lens, and with additional information and context that hadn't been told earlier. Besides, I felt these stories deserve to be told from different perspectives to reach a larger audience. The ten stories chosen for this book were determined by their extraordinary uniqueness, distinct protagonist voices, and context and relevance to the theme of the book, which represents forgotten stories from timelines of important points in India's history.

A century in three parts

Part I of the book begins with stories from WWI when 1.5 million Indian soldiers went to war fighting for Britain in the battlefields of France, Belgium and in the Middle East against Germany. This was the largest expeditionary force – bigger than the combined numbers from Scotland, Wales and Northern Ireland in the war. The readers might find it intriguing that neither were these men fighting for a free country of their own nor did they harbour enmity towards Germany or have any familiarity with Belgium or France. But the soldiers landed in unfamiliar territory, fought hard battles and won the hearts of the locals. In World War II (WWII), while most Indian Army soldiers fought on the side of the Allies against the fascist Nazis, some were faced with a testing predicament as Indian freedom fighter Netaji Subhas Bose's Indian National Army (INA) fought on the enemy's side in the war. There is a story in this book in which a protagonist, fighting in WWII, faces this dilemma. Stories of resilience and survival ran alongside a period of the country's history when Britain's political and economic needs were inextricably linked to the supply of manpower and raw materials from India.

The British belief in the supremacy of martial races led to large-scale army recruitment from among the Sikhs, Gorkhas, Garhwalis, Rajputs, Pathans and others. The soldiers mostly came from designated fighter classes from villages whereas officers were largely from among privileged royals. One of the stories in the book explores a different category of fighters – pioneers who came from a class that was neither the royal class (the kings and nawabs who mobilized their forces during the war) nor hardy rural folk, but an upwardly mobile professional class. These men were led by their ambition to fly in the skies. Laddie Roy and Hardit Malik were pioneering flyers who were obsessed with flying planes at a time when war pilots were known as the '20-minute club'. This was because, in 1916, the average life expectancy of a pilot in combat was twenty minutes. Stories about trailblazing Indians whose roles in tank and airplane combat – two of the most destructive vehicles in twentieth-century warfare – have remained largely unsung. One story involves the deep trauma of a prisoner of war, who fought a lonely battle all by himself. When Chanan Singh Dhillon went to serve with his unit in Africa in 1940, little did he know that he would spend the entire WWII with his captors, oscillating between life and death. While in prison, he wrote meticulous notes in a diary, preserved and shared by his children, which found its way into this book as a story.

Part II of the book traces stories of soldiers whose actions while defending India's territorial integrity reflected the fierce pride of a new, independent nation. In a story from 1948, a young Ladakhi boy volunteered to join the army to protect his land from invaders. Inspired by his ancestors, Chhewang Rinchen would go on to become a remarkable figure in India's military history. The indomitable Chhewang took part in every war that India fought: in 1948, 1962, 1965 and 1971, and returned undefeated each time.

India's leadership lapses and tragic reverses in the 1962 war with China were an ironic contrast to the dogged courage of its young officers and soldiers who unflinchingly put their lives on the line. The story of Haripal Kaushik helps us understand a phase of India's history when the country was dealing with a serious challenge from an emerging neighbour and simultaneously faced another bête noire across the border to its north. The story also features the effect of conflict on post-traumatic stress disorder (PTSD) and mental health. In this period India's borders were vulnerable, its leadership was rattled and its adversaries were posing a stiff challenge.

Nine years later, the contagious energy among fighter pilots in a story at the edge of the 1971 India–Pakistan war reveals a new, confident nation. India's turnaround was bolstered by a military that had acquired a definite national character but retained its ethos and fairness. As borders turned porous and geopolitical changes occurred in Afghanistan in the 1980s, the world began to face the twin issues of insurgencies and cross-border militancy. Adversarial governments in countries such as Pakistan began to fund terrorists, leading to a rise in violence in the Kashmir valley.

The stories in Part III of the book deal with complex modern era issues of internal security, the effect of conflict on entire generations who never had the chance to participate fully in life: lost generations, and the impact of violence on the people involved and their kin. This part highlights how the trauma of battlefields had shifted closer home, and conflicts began to be beamed live, affecting soldiers and their families. Insurgencies became a prominent feature in the 1990s, leading to faceless adversaries. In the nation's quest against rising militancy in Kashmir, soldiers in operational areas were the ones with ears to the ground as they sought to win back the hearts and minds of disaffected youths who faced difficult choices. Whilst we are familiar with events around the Kargil War, Kashmir militancy and the Mumbai terrorist attacks of 26/11, I was both surprised and fascinated by the unusual and powerful stories that lay concealed beneath well-known events.

Not heroism alone

Aside from gallantry and courage, the stories reveal the vulnerability and humaneness of the protagonists. Going beyond the binary of 'friend' and 'enemy', amid the fog of war, the narratives uncover interesting relationships between adversaries. There are instances of mutual recognition of abilities and duties. For instance, the story of Hardit Malik in WWI led to an unexpected reunion with an enemy six decades later. The story of daring pilots from the 1971 India–Pakistan war was centred around a fascinating encounter between adversaries who acknowledged each other's abilities many years later. In another story, when two foes were poised to enter a battle in the Himalayas and discovered more about each other, they invented a reason to save lives on both sides. In another story, an army captain was involved in an encounter with a militant where they kept firing at each other in between having conversations on the subjects of terrorism and jihad. In a ferocious battle with no quarters given, the two adversaries kept their cool and maintained decorum despite the strongest provocations.

Some stories in the book explore a human's quest for

freedom, longing for their homeland, while some others evoke incredulity. Stories of protagonists offer different perspectives into the landscape of war – where no experience is right or wrong, and hope and honour become central to survival and the aftermath. The stories pursue a fresh take on Indian military history and battlefield experiences that go beyond the stereotypes of 'gallantry and heroics' and explore a wider canvas.

Preserving stories

There are recorded, illustrious works of many exceptional writers and historians on war. I am grateful to historians such as Santanu Das, Shrabani Basu, Jagmohan Sapru, Kaushik Roy, George Morton-Jack and others whose remarkable writings on World Wars have helped acquire insight into the travails of Indian soldiers in the wars. Notes in the diary of Lieutenant Colonel Chanan Singh Dhillon, shared by his son Gurbinder, helped reconstruct his astonishing journey in WWII. In post-independence India, the works of Samir Chopra and P.V.S Jagan Mohan, Shiv Kunal Verma, Arjun Subramaniam, Claude Arpi, Gary J. Bass and many others helped understand the context of stories in this period. One of the more fascinating aspects during research was in listening to the experiences of soldiers involved in operations and events. These sources were not writers, but characters in their own stories, whose narratives – candid, intuitive and first-hand – provided irreplaceable, original and rare insights. The research on this book began during Covid-19 when travel was limited. Thereafter, interviews and meetings took me to Punjab, Leh and the Galwan region in Ladakh, the Kashmir valley, the East Khasi hills in Meghalaya, Kalimpong, Pune, Kumaon in Uttaranchal, Karnataka among others. During the course of writing this book, apart from many hours of research and gathering information from multiple secondary sources, I interviewed 40 sources spending around 400 hours on discussions, with multiple extensive sessions and phone calls with several of them in order to reconstruct their stories.

The purpose of the book is to contribute to the preservation of the perspectives of protagonists that would be otherwise lost over time. In Britain, the Imperial War Museum's archives have a war diary from the battle of Cambrai with handwritten notes scribbled about the time, nature and place of actions involving Lance Daffadar Gobind Singh and his unit. A gate-pass for an Indian soldier signed by the Nazi supervisor of a German prisoner camp in WWII survives with the family who have carefully archived memories from the war. A letter written by the Pakistani air chief to an Indian Air Force officer many years after the former was beaten in an air battle in 1971 is a piece of rare history presently in possession of his rival: the Indian Air Force pilot. This book can be seen in this tradition of preservation and taking these stories to a general readership.

Another aim of the book is to support a growing interest in the history of the Indian Army through the medium of short stories. In 1915, Indian soldiers recorded the only victory of the Allies in the Gallipoli campaign against the Turks. Today, the Gallipoli campaign is commemorated on ANZAC day, a national day of remembrance, each year in Australia and New Zealand because their troops fought in Gallipoli. Sikh, Punjabi and Gurkha troops who fought in the same campaign have sadly been forgotten. Memories, weapons and items from the Gallipoli campaign are carefully preserved in the Australian War Memorial in Canberra, which I happened to visit. But we have none of this specific memorialization in India beyond the war memorial at India Gate in New Delhi which commemorates generally the sacrifices of Indians who died in the World Wars and the National War Memorial, built in 2019, which honours soldiers who gave up their lives fighting for independent India.

Despite the fact that several thousand Indians participated in the two world wars, rarely have Indian characters been featured in Hollywood's war films. In fact, the story of a recent Hollywood film bore uncanny similarities with one of the stories in the book about an Indian war hero. The Hollywood film went on to win three Oscars whereas the Indian hero of Cambrai has remained largely unsung, except for being celebrated as a legend by his regiment.

India ought to have more war stories than any other nation. The last hundred years of history have witnessed significant changes in political and geographical identities that impact soldiers, their families and others involved in these events. These experiences reveal a range of emotions including joy, pain, survival and sacrifice that have long lain hidden in cryptic notes of war diaries, in regimental chronicles or in the anecdotes of military veterans. It is this collection of forgotten, lesser-known stories that this book, *Camouflaged*, brings to you.

Part I

Indian Soldiers, Foreign Wars

In World War I (WWI) and World War II (WWII), Indian soldiers embarked on journeys across continents, enduring long separations from their families while fighting in wars that were not their own. Many of them faced years of imprisonment or died without recognition, and their tales of bravery and survival remained unknown.

These chapters showcase the hidden heroes, from pioneers and adventurers to prisoners of war (POWs). Their stories tell us about a time when war was far away from home. In unknown lands.

1

Sultans of the Sky

Hardit Singh Malik and Indra Lal 'Laddie' Roy

On 17 December 1903, Orville and Wilbur Wright flew the first aircraft at Kill Devil Hills, North Carolina, with Orville piloting the plane that they had invented, called the Wright Flyer. Eleven years later, their invention emerged as the world's foremost killing machine during World War I (WWI): bombs rained down from airplanes during the conflict.

And thus, the modern air force was born.

By 1918, WWI had been raging for four long years. The violent duels in the air – close-range aerial battles that took place between opposing fighter planes, whose pilots engaged in combat manoeuvres and tried to shoot down enemy aircraft – kept the people down below on tenterhooks, as these dogfights often ended in plane crashes leading to the destruction of their land and homes. The fear of the sky falling on them had never been more real, and they lived in constant anticipation of the next devastating event.

On 22 July 1918, villagers in Carvin in western France

awoke to a familiar sight. Four German Fokker D.VII fighter planes were tussling with a lone British aircraft. Ground troops on the front, including an Indian gunner, were riveted to the unequal contest that was unfolding. The British fighter was hit by enemy fire and suffered damage to the fuselage (main body) of his plane. Yet, the fearless pilot emerged as the underdog winner of the contest, taking down two from the enemy pack with aerial gunfire in a final hurrah before plunging into flames.

After the war, a majestic monument was erected for the young fighter of that battle in the Estevelles Communal Cemetery in rural France. His name was Indra Lal 'Laddie' Roy.

Few persons who have exhibited such extraordinary bravery have died so far from their motherland. The epitaph on Laddie's tomb was in two languages – French and Bengali. 'Mahabirer samadhi; sambhram dekhao, sparsho koro na . . .' (The grave of a brave warrior; show respect, don't touch it . . .)

While much has been said about fighter planes revolutionizing warfare during WWI, the valiant Indian fighter pilots who ventured into the heart of battle have remained largely unknown. Fearless flyer Laddie Roy was following in the footsteps of Hardit Singh Malik, the first Indian fighter pilot.

This is their story.

On 1 June 1912, author Sir Arthur Conan Doyle,[1] playing for the Sussex Martlets, stepped up to bowl to a stylish Indian batsman from an Eastbourne cricket team, from the town of East Sussex, UK. During that match, the Indian batsman scored 19 runs out of a total of 85[2] before being stumped by Doyle's wily off-spin delivery.[3] His impressive batting skills caught the attention of a friend of Ranjitsinhji, the Indian cricketer, who recommended the teenager, Hardit Singh Malik, to Sussex cricket team captain Herbert Chaplin.[4] By 1914, Hardit was playing for the Sussex County in the English cricket league. However, his future was about to take a dramatic turn, changed by an event in Sarajevo.

On 28 June 1914, the assassination of Archduke Franz Ferdinand, the heir presumptive of the Austro-Hungarian throne, and his wife Sophie, by a Bosnian revolutionary in Sarajevo, had led to the outbreak of war. As Germany, Russia and France issued ultimatums to each other, the long shadow of war fell upon Great Britain which would soon join the conflict.

Hardit, who had been sent to England from India to study at the age of 14, was at Balliol College, Oxford, and paving new paths on the cricket field when war broke out. When he took the field on a rainy day in August that year, playing against Kent, the future of the cricket season was uncertain. Inclement weather interrupted play at Canterbury. During the rain break, Hardit strolled across the ground with his friend, Kenneth Woodroffe. Their chat gravitated towards the war. Woodroffe declared he would join the army and fight if and when Britain joined the war. He then asked Hardit what he would do. The Indian smiled, shrugged his shoulders and ambled towards the pavilion.

That was the day Great Britain entered the war.

Ten days later, Woodroffe joined the 6th Battalion, Rifle

Hardit played five County matches for Sussex in 1914.

Brigade. Three weeks later, the county cricket season was called off due to the war. Within a year, Woodroffe was killed in the Battle of Neuve Chapelle.[5] Hardit eventually went to war too. However, unlike Woodroffe, his journey to the frontlines took a different path.

WWI witnessed the widespread use of trenches in battlefields and drew infantry soldiers from various countries, including over a million from India, all fighting under the banner of the British Empire. The era marked a shift in warfare, as newly invented potent forms of grenades, artillery and explosives decided the fates of soldiers skilled in close-quarter battles with rifles and bayonets. And then came the aircraft, whose introduction brought about a radical transformation in the war.

Hardit

Flying had always fired Hardit's imagination. He even had a premonition about his destiny, as a child.

Born in 1894 in Rawalpindi, west Punjab, he grew up in a privileged home and enjoyed an idyllic childhood. Flying kites was a beloved activity, and he would join the other boys in launching them across terraces and gardens. In a sky speckled with colourful paper kites, Hardit delighted in their

Hardit Malik: a portrait

leisurely flight and eagerly joined the kite duels, revelling in the victories as defeated kites gently descended to the ground.

Kites fuelled his dreams. *Would he be on a flying machine in the sky, some day?*

At Oxford, the choice wasn't easy. To fulfil his passion for flying, Hardit would have to fight for a colonial power that ruled his country by force. Though, ironically, he would be participating in a war against an expansionist power, as part of the Allied Forces (including countries such as Great Britain, France, Russia and later the US) opposing the Central Powers (including Germany, Austria-Hungary and the Ottoman Empire).

Watching his peers at Oxford head to the frontlines, Hardit applied for a fighter pilot's position at the Royal Flying Corps (RFC). The prospect of an Indian in that role raised a few sniggers, and his application was rejected. Undaunted, Hardit urged Francis Urquhart, his Oxford tutor, to recommend him for a civilian support role in France.[6]

After graduating in 1915, Hardit was hired by the French Red Cross to drive a motor ambulance to the frontlines. It was an uneventful assignment until fate intervened unexpectedly.

While driving through the picturesque countryside, Hardit would witness formations of airplanes zooming overhead. Watching those planes triggered memories of his childhood days, that were spent flying kites and another powerful realization: his true passion was flying. He couldn't let this opportunity slip through his grasp.

He asked a friend in Cognac if he could apply to the Aéronautique Militaire (French Air Service) and, to his surprise, they accepted his application. Filled with excitement, Hardit wrote to Urquhart, who was far from elated to hear that a British resident was to join the French Air Force. Urquhart, a respected academic, shot off a terse letter to Major General David Henderson, chief of the RFC. He wrote, if Hardit Singh Malik as a British subject was good enough for the French, why wasn't he good enough for the British Armed Forces?[7] The letter worked, and a few days later, Hardit was summoned to England and found himself facing General Henderson in the latter's office. On 5 April 1917, he was fast-tracked into service as Honourable Second Lieutenant H.S. Malik, RFC, Special Reserve.

Hardit's resolute push for what he wanted didn't end there. As a cadet at the prominent military training academy in Aldershot, he stood out as a devout Sikh, proudly wearing a beard and turban. At the time, British Army regulations mandated clean-shaven appearances and prohibited headgear, to ensure the effectiveness of protective equipment and uniform conformity. The sergeant major was aghast at this departure from regulations, but Hardit stood firm on religious grounds. The matter was taken to the commanding officer (CO), who faced a difficult decision as there was no prior occurrence. For that matter, the induction of Indians into the air force also lacked precedent.

Eventually, the demands of war required the setting aside of colonial egos. Hardit was given a special helmet made to fit over his turban. His unique appearance earned him the nickname 'Flying Hobgoblin' at flying school!

Hardit's incredible story finally took wing. Action awaited him in the skies.

Tales of soldiers from the frontlines inspired young men to join the war. However, until then, no one had heard of Indian fighter pilots. Lieutenant Jeejeebhoy Piroshaw Bomanjee had become the RFC's first Indian pilot,[8] but he resigned due to ill-health.[9] After him, as Hardit broke barriers and opened doors, other Indians started to follow in his footsteps.

Unlike the British Indian Army, the air force was devoid of any legacy of 'martial races' policy that enabled the British to exclude the educated populace from military service.[10] (During the colonial era, the British classified specific ethnic groups or communities as having inherent martial qualities and gave them priority in military recruitment. While the Sikhs, Rajputs, Pathans and Gurkhas were perceived as loyal and possessing fighting qualities, certain regional groups were thought to be dangerous and disloyal to the British, and thus were considered unfit for the military.)

The RFC played a role in shaping a new class of fighters – privileged Indians. Errol Suvo Chunder Sen and Shirikrishna Chunda Welinkar, hailing from well-heeled families, would later join Hardit as fighter pilots in the war.

As did a third pilot who, having just completed his training, was about to script a sensational story in the skies.

Laddie

On the sunny afternoon of 28 October 1917, Major Rainsford Balcombe-Brown, CO, 56 Squadron, looked at his watch. With an hour until the evening parade, he reclined on his couch, engrossed in a two-week-old edition of *The Times*.

A young Indian boy, fresh from flying school, entered the room.

'How old are you?' Brown asked.[11]

'Nearly 20, sir,' the boy said. (He was 18.)

'Get plenty of flying in, because you will need all your skills,' was Brown's advice. He added, 'I suppose you are too young to think and too old to listen. Are you valiant, gallant and dashing?'

The question was met with a quiet affirmation.

Brown seemed unconvinced. 'Why are you in this war?'

'Sir, I don't understand.'

'You are an Indian prince. Why should you be in this war? No one has compelled you and conscription does not apply to you.' (He wasn't actually a prince but came from a prominent zamindar [landowner] family.)

The boy paused for a moment. 'I volunteered, sir.'

Brown tried to dissuade him. 'I suppose you wanted adventure . . . do you know it's a killing field? You have to survive and fight. And then survive again.'

'I came prepared for that, sir,' the boy said calmly.

'Welcome aboard,' Brown shot back with a smile.

And that's how Indra Lal 'Laddie' Roy arrived as a fighter pilot and became a part of the war.

Years ago, in 1901, Lolita Ray, wife of Piera Lal Roy, an eminent barrister from Calcutta (now Kolkata), arrived in England with her children, hoping to ensure a proper English education for them. The family had its roots in Barisal (now part of Bangladesh), a region known for the fiery temper and raw courage of its inhabitants.

The boys, who were sent to St Paul's School for boys in

Laddie at school

Hammersmith, were of a sporty temperament. Paresh, the elder, was the school's boxing champion while Laddie played rugby and captained the swimming team.[12]

When news of war broke, the boys – who had grown up to be true Englishmen[13] – immediately expressed their desire to enlist. Paresh had just graduated from Cambridge and was 'keen to prove that the non-martial Bengali could fight as well as any other soldier'[14] while Laddie, who had won a scholarship to Oxford, dreamt of flying. The heroic exploits of Albert Ball, VC, renowned British flying ace during WWI, kept him awake. Laddie spent long hours by the window, immersed in thoughts about the new winged machines that now ruled the sky. He wanted to be a fighter pilot, like Ball.

He got off to a disappointing start. Laddie's application

to the RFC was rejected on grounds of poor eyesight. With unwavering determination, he sold his motorbike to afford a second opinion from a renowned eye specialist in the country.[15] Laddie successfully passed the eye test, leading to the reversal of the decision, and on 4 April 1917, he enlisted in the RFC. Paresh joined the reserve battalion of the upper-class Honourable Artillery Company.[16]

Their mother, Lolita, was proud that Paresh and Laddie had both enlisted, but she could not help her tears as young Laddie was about to leave home.[17] 'My brave *Barisal Guns* of the war,' she'd call them, referring to the unexplained sonic booms – 'skyquakes' – that rocked Bangladesh's southern coast.[18]

Laddie's selection was a celebrated event in the family. His sister Leila promised to buy an RFC sweetheart brooch, a pin depicting the regimental affiliation of a serviceman, usually worn by their wives or girlfriends. Emily, the girl he had courted despite her initial rebuffs, had been supportive too.

'Will you write to me?' he asked.[19]

'Of course, I will,' she said.

After a period of flying training at the Royal Naval Air Service flying training school at Vendôme in central France and a spell at the RFC Gunnery School in Turnberry, Laddie was given his first posting by October 1917. It was with the 56 Squadron, the unit of his boyhood hero Albert Ball, VC. By the time Laddie joined the squadron, which was stationed at Estrée-Blanche airfield near the town of Étaples in France, Ball had already been killed.[20] Within the space of three months, Laddie had earned promotion to the rank of second lieutenant.[21]

When Major Balcombe-Brown saw Laddie in his tent at Vendôme, he noticed that beneath Laddie's gentle demeanour, there resided a passionate spirit that aspired to bring down German planes. Two months later, Laddie's time would come. However, it wasn't quite the performance Brown had expected.

※

The war had claimed a large number of young pilots. As a result, after a shortened period of training, cadets were quickly drafted into frontline combat within months of their instruction (see map on page 15).

Soon after his commission, Laddie was flying over the frontlines in France. The war was at its fiery peak with an aggressive Luftstreitkräfte (Imperial German Army Air Service) targeting British planes.

In one of the sorties, Laddie's S.E.5 aircraft was picked up by three German planes. They homed in and unleashed a flurry of fire, catching Laddie by surprise and smashing the control column (joystick) of his plane. The Indian dived instinctively, losing height to throw them off his tail. But the three planes were in hot pursuit. Laddie's aircraft, battered by the attacking planes, crashed.

When his plane hit the ground, Laddie was found strapped to his seat, bloodied and unconscious.

He was taken to a local hospital near Estrée-Blache. The hospital placed him in a morgue, but neglected to ascertain whether the pilot may still have been alive. So, in the middle of the night, when Laddie came to his senses, he found himself in the company of the dead. That night, Rene Fonck, the morgue attendant, heard noises on the door of the mortuary. Curious, he crept up close to the room with a lantern. He heard someone banging on the door and asking for help in

basic French. The terrified attendant froze with fright. 'Who is it?' his voice quivered.

'It's me... Lt Roy. Please let me out. I am not dead... but hurt and mistaken as dead.'

It was midnight when Fonck opened the door after much hesitation. Before him stood Laddie, bloodied with multiple injuries, in a state of delirium. Fonck rushed him to a doctor, attended to him and after a week, put him on a hospital train leaving for the coast.[22]

Needless to say, it was a stark initiation into the harsh realities of war for the teenager.

Three days later, a shaken Laddie found his way from Calais to the shores of England, having survived his first tryst with combat. A long, painful period of recovery awaited him in England.

Hardit

After Aldershot, Hardit honed his skills further at Fulton where he flew combat planes including the Avro 504, the Sopwith Pup, and finally, the Sopwith Camel, a single-seat biplane and the most advanced fighter at that time. There, Hardit picked up various combat tactics, such as the tricky Immelmann Turn, an aerial manoeuvre that involves a swift dive, climb and loop. After his time at Aldershot and Reading, he was sent to Vendôme in France where he flew his first solo after just three hours of flying.[23] A quick learner, Hardit got his wings in under a month.

In October 1917, he was assigned to the 28 Squadron in France and equipped with the Sopwith Camel. (Coincidentally, Laddie's 56 Squadron, also called the Punjab Squadron, was formed out of the 28 Squadron.) As the war in France intensified, the formation located to an airfield near the village of Droglandt in Flanders.

Hardit found a mentor at the squadron in his flight commander Major William 'Billy' Barker, a Canadian who had joined the RFC in 1916. A skilful fighter, Barker was considered the greatest all-round pilot of WWI and would go on to win the Victoria Cross for gallantry.

He personally initiated Hardit into the art and science of aerial combat.[24]

Hardit's first flight over the German lines took place on 18 October and was relatively uneventful. On 19 October, when the 28 Squadron went into action alongside 70 and 23 Squadrons to make a combined attack against the important German aerodrome at Rumbeke, a Belgian village near the town of Roulers, Hardit was part of the task force under the command of Captain Barker. The 28 Squadron's brief was to surprise and engage the German fighters dispatched to intercept the bombers. At first, it was thrilling to be part of such a large formation. Hardit began to soak in the experience of flying close to enemy lines. Soon, however, the situation took a chaotic turn. Barker, flying alongside, gestured to him. A group of German planes were heading straight towards Hardit.

In his account: 'There were bullets flying in all directions. We had been instructed that each pilot was to pick out one particular target, and I soon found myself diving at the tail of an enemy who, instead of turning back to attack me, kept on diving. He must have been as frightened as I was! I must have started shooting from too great a distance, for at first nothing seemed to happen. But suddenly I hit him and first his plane started to smoke, and then went down spinning in flames.'[25]

Flying too low, Hardit needed to swiftly ascend to avoid a similar fate. Luckily, Barker noticed his predicament and guided the task force to assist him. They climbed to a safer altitude and set course for Droglandt.

Barker led Hardit in his first few actions, giving the young man his early experience of ruthless duels. Hardit was conspicuous by his special helmet and Barker fondly called him 'Indian Prince'. Determined to stamp his country's presence in the war, despite officially representing Britain, Hardit proudly had 'India' written on the side of his aircraft. In the weeks that followed, he achieved eight more aerial victories, demonstrating his exceptional skill.[26]

Hardit was now a hero.

Laddie

By January 1918, while Hardit was making a mark for himself, Laddie lay in a hospital bed in England nursing a broken shoulder, fractured leg and, among other injuries, a bruised ego as well.

The war had separated him from his loved ones but now, in a weakened state, he felt alone and longed for his family. Deeply attached to his siblings, he often thought of his mother Lolita, and brother Paresh, a gunner in the army who was serving at the front. He missed his mother the most and hoped to make her proud. To pass the time, he sketched fighter planes.

On his bed, Laddie was drawing the Dreidecker ('triplane' in German) that belonged to German flying ace Captain Manfred Albrecht Freiherr von Richthofen. A thorn in their side, the British had nicknamed him 'Red Baron'. (He earned the name 'Red' due to flying a red-painted Albatros aircraft, while 'Baron' was derived from his German title 'Freiherr'.)

Just then, there was a faint knock on the door.

'Hello Laddie.'[27]

'Hello Emily! How did you know I was here?' Laddie sounded happy she was keeping track of him in the war.

'Your mother told me, and here I am.'

Emily's arrival was like a burst of sunshine. She was his love, confidante and a link to his family.

She sat near him, looked deeply into his eyes and took his hand in hers. He shut his eyes as she ran her fingers gently through his hair. For a moment, to her, he appeared to be the young lad in his final year at St Paul's School who had vented his frustrations at her for being rejected by the air force.

'Everyone is fine and sends their love to you. she assured him your brother is at the front . . .'

After a while, she said, 'Darling, I have to go now. I promise I shall come every day to see you.' Emily hadn't written to him, though she had promised earlier. Her eyes turned moist and with their hands clasped together, she bent over and kissed him. 'You are the only one I love,' she said, her voice faltering as tears rolled down her cheeks. Laddie pulled her face down towards him, holding her lips close to his. He could smell the warm fragrance of her hair as the locks fell all over his face, burying his eyes and ears in her tresses. For a moment, he wanted them to be in this embrace forever. When she got up to go, she held his hands and kissed them lightly.[28]

Between Emily's visits and his sketching, Laddie's time in the hospital flew by. By fall, he was discharged. He went with Emily to London and met his family.[29] It was an idyllic time; they went to the theatre, restaurants and took long walks by

Sketch of a Sopwith Camel drawn by Laddie

the waterside. One evening, they went to watch the musical hit *Fair and Warmer*, a favourite with the two.[30]

The joys of city life were cut short when Laddie's posting order arrived. It wasn't happy news. He had been declared as not fit to fly and was being sent to the 40 Squadron.

Laddie held his head in his hands and sank on to his bed. 'I am now an Equipment Officer . . . imagine me being on the ground while the other fellows are up in the air,' he ranted to Emily. 'He is a beautiful, natural pilot but tends to get excited and leave the formation,' read the assessment report signed by Brown, his CO who Laddie had impressed earlier.

Emily smiled gently and tried to console him. Laddie's heart sank, but he was determined to prove his seniors wrong. On 19 June 1918, he returned to duty with 40 Squadron based at Bryas in France. Although posted to the squadron as an

Equipment Officer, within days, he had persuaded the medics to pass him as fit for flying duties.

For a warrior to redeem himself, skills and courage alone may not be enough. Having a mentor increases the chances of their abilities flourishing. At 40 Squadron, Captain George 'McIrish' McElroy took an early interest in the troubled young pilot and gave Laddie permission to fly once he was cleared by the medics.

One morning, McElroy and Laddie were piloting in an area around Drocourt in northern France, when the latter saw two approaching German aircraft and dived into them. Nimbly dodging the German gunners on the ground, Laddie kept one of the aircraft in his sights. When it was 80–100 yards away, he pulled the trigger and unleashed a burst. To his surprise, a barrage of rounds crashed into the German aircraft making it lurch and crash. McElroy downed the second German aircraft.

When they landed, there was jubilation all around about Laddie's performance. 'From a bumbling beginner to an ace in the making,' remarked Reed Landis, one of the ace pilots at the base. Laddie had planned to retire to his billet and work on a special sketch of the first German plane he had shot down when McElroy stopped him. 'Hey! It is evening, and all good Irishmen celebrate. So, Laddie, off we go to the mess.'

Amid loud cheers from the officers, they headed to the bar. 'Beer,' Laddie ordered promptly, raising a toast to the man who had bloodied him in combat earlier that morning. Like Hardit had Barker, Laddie had also found a mentor in McElroy. The captain wore the look of a proud teacher whose pupil had bested the odds.[32]

On that day, 6 July, Laddie had shot down his first German

aircraft,[33] but there were more victories to follow. 'You will fly again and become a famous scout pilot, darling, mark my words,' Emily had said to a downcast Laddie before he left for 40 Squadron. He had begun to prove her right. Moreover, that morning, in his protégé, 'McIrish' McElroy had seen the future.

Hardit

Laddie's sketches included one of the plane flown by the Red Baron, who had acquired iconic status by downing over 80 British aircraft in dogfights.[34] The Baron's squadron, Jagdgeschwader 1 (JG1), was referred to as the 'flying circus'[35] due to their colourful aircraft and distinctive style of aerial combat. They were known to add an artistic flair to their dangerous missions.

Hardit Singh Malik fought the Baron's circus too.

It was during the Battle of Passchendaele, a brutal and costly WWI battle fought in the eponymous Belgium village from July to November 1917, that Barker heard about the presence of the Red Baron and his circus across the lines in the nearby village of Marckebeke. He moistened his lips; this was the moment to take on the German aces in another duel. Sidestepping his cautious boss, he obtained permission to attack the Germans and chose a crack team of skilled pilots for combat. Including Barker, the roster consisted of Lieutenants N.C. Jones, J.B. Fenton and Hardit Singh Malik.

On 26 October 1917, they set off towards the village of Poelcapelle, an area in Belgium, heavily affected by battle. It was a dark and wet day; it had been raining all night in Poelcapelle and there was no sign of the showers abating.

Visibility during the day was poor. To fly was dangerous, let alone duel in the sky. But Barker had other ideas. The presence of Hardit Singh Malik reassured him, as he knew the latter was driven to succeed.

Barker was excited at the prospect of taking on the ace fighters of Jasta 18, one of the fighter squadrons that composed JG 1. This was his shot at making history!

On the soggy RFC field at Droglandt, Hardit looked up from his Sopwith Camel B5406, wondering if it was the most reckless decision to fly that day. At 1045 hours, the ground crew watched as four planes struggled to lift off, navigating through a soft drizzle and disappearing into the darkness.

Around the same time, four Albatros fighter planes took off from Harelbeke, a German-occupied town in West Flanders, Belgium, and headed into the showery, robust southwest wind. The German fighters were manned by four of the most seasoned veterans of Jasta 18 – Paul Strähle, Otto Schober, Arthur Rahn and Johannes Klein – who were ready to take on the British fighters. It was going to be a busy day.

Like Hardit and others on the Allied side were inspired by the early British pilots, young German men were drawn to ruling the skies spurred by the heroic deeds of the Red Baron. Paul Strähle was one of them.

In Brussels when war broke out, Strähle hoped to join the Luftstreitkräfte since he had trained in piloting a Zeppelin, a large, cylindrical shape airship. Initially assigned to the infantry, he endured the dangers of trench warfare before transferring to aviation. He flew his first combat mission as a pilot at Ypres, West Flanders, in mid-July 1917. After receiving his pilot badge in August, he rose up the pecking order. Posted to Jasta 18, Strähle would soon fly over Flanders.

British pilots played a pivotal role in the aerial battles of this region, and Strähle eagerly awaited the opportunity to engage them. Like most German fighter pilots, he felt that the RFC gave him more chances to test his combat flying skills; the RFC was aggressive and unrelenting, unlike the more defensive French.

Strähle and Hardit were about to be bound by their destinies to the war.

Barker, Hardit, Jones and Fenton started out together but shortly thereafter, owing to poor visibility, Jones was separated from Barker and so was Fenton. Hardit, though, stayed close to Barker's B6313, barely managing to keep him in his sights in the bad weather.

West of Roulers, a town located between Marckebeke and Ypres, the German Albatros formation flew into the paths of the Camels of Barker and Hardit. Strähle saw two single-seaters in the far distance; Barker and Hardit were about 1,200 metres away and heading towards him.

Hardit swerved from the path and dived down to engage German targets on ground. Strähle decided to chase him, and Rahn followed. Klein and Schober now teamed up to engage Barker. The two British pilots had been outnumbered by the Germans.

The weather wasn't making it easy, but Hardit managed to shoot Rahn who gave up and tried to land his aircraft.

Strähle, though, was unflappable and determined to hang on to Hardit's Camel. The Indian realized he was being tailed and tried to weave out of the way, pulling out different tactics he knew of. The dogfight got intense as both ace pilots tried to evade each other, flipping their machines and changing directions to get a good shot at the rival. They dived down and swung sideways to avoid being targeted, and rapidly kept losing height. At one time they were just a few feet off the ground, charging desperately at each other. They pulled back and climbed again – swerving in the skies like two ace motorists in the air, manoeuvring deftly to avoid death and briefly gaining an advantageous position over the other.

And then the tables would be turned again.

Testing each other's skills and patience, hoping the other would make a mistake, the dogfight lasted more than a quarter of an hour.

Hardit managed to hit Strähle and saw smoke coming out from his aircraft. A moment later, the former experienced a sharp, metallic pain. Two of Strähle's bullets had hit him. The pain began to spread to Hardit's thighs and waist, and he could see his bloodied trousers. His head throbbed relentlessly, yet he held on.

On his tail, the Germans fired relentlessly – over 400 bullets – at his aircraft. Hardit tried to fight the pain, the enemy and the weather as he looked for an escape plan. The cockpit grew increasingly hot as the petrol tank had been hit. His mind began to fade into a peculiar numbness. Was he losing consciousness? Would he end up crashing?

In the distance, in the sky, he could see Barker being surrounded by the Germans. That was the last he saw of Barker on that day.

Hardit dove downwards as he struggled to keep the aircraft steady. His plane stuttered and swayed sideways as it rapidly lost height. Where could he land his Camel? The clouds had disappeared and just one spot of swampy land was visible. The searing pain in his leg was unbearable by now. Should he risk the swamp? A soft surface would absorb the velocity of impact and maybe save him. Hardit landed inside the French territory, skidding along the mud-spattered surface, swaying and tilting dangerously till he lurched to a screeching halt.

One of the medical orderlies who took him out on a stretcher later recounted it was a miracle Hardit had survived. He had lost a lot of blood, broken his nose and taken bullets. He had, however, retained his indomitable spirit.

Doctors advised Hardit to leave the lodged bullet in his thigh, as attempting its extraction posed greater risks than living with it. During his hospital recovery, he wrote a report detailing the operation and expressed his concern for Barker,

Hardit's Medical Casualty Card after being wounded

whom he had last seen surrounded by German airplanes, unsure if he would survive the encounter.

Incidentally, Barker did survive. Ironically, his report was identical to Hardit's except he said he didn't think the Indian would survive!

Both were unaware that they would live to fly again together in the war.

Laddie

By July 1918, Laddie had overcome his early setbacks and become a glorious veteran of war, a teen sensation who rapidly gained an envious reputation as an audacious fighter pilot.

Between 9 and 19 July, in less than a fortnight, he took down nine German aircrafts – a record at that time (though in the books he has been credited with eight victories).[36] He would return to the base and draw sketches of the German aircraft he had fought.

In the air force messes and amongst his service colleagues, Laddie had become a venerated figure for his devil-may-care attitude. They spoke reverently of a brown man who had turned a white man's invention into his machine. Most pilots were in awe of this young Indian prodigy. He had become known among the Germans as a fearless fighter, evoking both their hatred and admiration for his sheer courage. His brisk rise meant he was going to be targeted by the enemy soon.

On 22 July 1918, Laddie took off for a patrol with two other S.E.5 fighter aircraft.

An uneventful forty-five minutes passed as he flew over the French countryside. The calm sky seemed unusual, comforting and eerie at the same time. Laddie had become a marked man. Were they setting him up?

He was above the town of Carvin in northern France when he noticed tiny dots in the far distance. In a short while, Laddie realized a formation of Fokker D.VI fighter planes was 100 feet above him.[37] (According to the CO, four enemy aircrafts attacked them.[38])

The four Germans engaged with Laddie, knowing well that they were probably targeting one of the prized fighter pilots of the war. Swooping down, they surrounded him as they tried to get a clear shot. Not one to give up, Laddie engaged with the formation. The young Indian was an agile flyer and his reflexes helped him steer clear, gain height and attack the enemies. The Germans had cut him off from the two other S.E.5s, one of which was being flown by Landis, and now circled him. It was an unequal contest, but he was more than equal to the task. He boldly attacked the Germans, rattling them.

Laddie's plane had been struck in the hull, but he dived along with the German Fokker before him. After the impact, he lost speed, but the Fokkers continued chasing him, wanting to ensure his end. When one of the German planes closed in on him, Laddie seized the opportunity. He manoeuvred and aligned himself with the German aircraft, closing the distance to approximately 50 metres. 'Now,' Laddie thought, and unleashed a burst of fire.

The Fokker was hit. Laddie continued firing bursts at him. The German plane dived and crashed to the ground.

He shot down two of the aircraft, but in the skirmish, while trying to get a clear line of fire on the third one, his aircraft was hit again. Laddie lost control of his plane. He looked at his cracked windshield. The earth seemed to grow bigger in a flash. Suddenly, a whirlwind numbed him. The next moment, it was all over. Laddie's plane crashed in the farm fields of Carvin.

A British Ack-Ack battery, a military unit equipped with anti-aircraft guns for defence against enemy aircraft, watched the contest along with the villagers of Carvin. They saw the British plane go down over the German lines. Among them was Laddie's brother Paresh. Three days before the air duel, Laddie had met Paresh and promised to shower flowers as he flew over him. That never happened.[39]

Laddie had crashed in German-occupied territory that morning. The German infantry ran forward to extricate him, but he was already dead. Indra Lal 'Laddie' Roy was given a military funeral by the Luftstreitkräfte, with a French priest conducting the burial service. He was laid to rest where he crashed. After the war, when his mother was asked if she would like the body exhumed for burial in England or India, she said that her son had died fighting and it was fitting that his body was buried at a place where he loved to be – the battlefield.

Laddie's CO wrote to his mother, 'From the time your son came to the squadron, his one aim was to shoot down Huns . . .'[40] There were pilots in WWI who shot down more aircraft than Laddie, but unlike him, no one else destroyed more enemy aircraft in a dizzying span of 170 hours. He was a trailblazer in combat flying in India.

At 19, Laddie had become the 'boy hero' of the war. He had proved Emily right. He had become a successful scout pilot. And a famous one.

He had made a will in which he left his money to his mother. If she didn't want it, it was to be given to Emily.

The end of the war

The armistice ended WWI on 11 November 1918 – three weeks before Laddie would have turned 20.

By the end of the war in 1918, approximately 1.3 million Indian soldiers fought in theatres across Europe, the Middle East and East Africa and more than 74,000 of them lost their lives.[41] India also contributed over $20 billion in today's money to the war effort, including 3.7 million tonnes of supplies and 1.7 lakh animals.[42]

The long war provided young Indian men with the platform to showcase their daredevilry. They battled monumental odds to become fighter pilots and they performed valiantly.

Laddie was posthumously awarded the Distinguished Flying Cross (DFC) in September 1918.

He became the first Indian to win the DFC. He was also the only Indian Flying Ace of the World War – a distinction never emulated again. A flying ace is a fighter pilot credited with shooting down five or more enemy aircraft.

His citation read: 'A very gallant and determined officer, who in 13 days accounted for nine enemy machines. In these several engagements he has displayed remarkable skill and daring, on more than one occasion accounting for two machines in one patrol.'[43]

After recovering, Hardit returned to the war and was posted to France with the Royal Air Force (renamed in 1918). He flew Bristol Fighters with the 11 Squadron until the war's end.[44]

He was the only Indian fighter pilot who survived the war and returned without being imprisoned.

It was by a strange coincidence that the ship Hardit boarded at Marseilles to return to India was the very same that he had travelled on in 1908 when making his way to England.

In India, Hardit went on to become a successful diplomat, appointed as India's ambassador to Canada and France. The memory of war continued to be lodged in the form of a bullet in his thigh, which his German rival Paul Strähle had shot at him.

Five months after the end of WWI, Hardit married his fiancée, Prakash. The ceremony took place on 13 April, the same day as Baisakhi, a spring harvest celebration in North India. Unfortunately, the festivities turned into a tragedy when news spread that Brigadier General Reginald Dyer had ordered his soldiers to open fire, resulting in the deaths of hundreds of innocent Indians at Jallianwala Bagh in Amritsar, which was the home state of Hardit.

He was shocked to hear about the killings and called it 'the most tragic, bloodiest day'.[45] Disturbances erupted, leading to the British Air Force bombing the cities of Amritsar and Gujranwala (now in Pakistan). They utilized Sopwith Camel aircraft, the very same planes that Hardit had once proudly flown in WWI, risking his life.[46]

Many years later, in the 1980s, Jagmohan Sapru, then a young journalist, began work on the biography of Ambassador Hardit Singh Malik, India's octogenarian diplomat statesman and a revered figure. While developing his narrative, he stumbled upon something incredible.

Sapru came to Hardit and told him that there was someone he would like to introduce to him. Having fought and survived the war, lived through the early years of the nation, rising to a high-ranking official and, in between, playing good cricket too, Hardit couldn't quite guess who it might be.

Was it a cricketer? Diplomat? Soldier?

Sapru meant it to be a surprise. Finally, on the appointed day, Sapru put Hardit in touch with the man. On the other side of the phone line was a person whose voice he had never heard before, but whose story was entwined with his.[47] It was Paul Strähle, the German pilot. He, too, had survived the duel and the war. While reminiscing, Malik learned that Strähle's guns had jammed, which resulted in them both aborting the duel. Sixty years after Hardit and Strähle had fought desperately to kill each other, fate brought them in touch.[48]

At the age of 91, Hardit Singh Malik, the cricketer turned prince-of-the-skies, died. He carried with him Strähle's gift to his funeral pyre – a bullet in his thigh.

Postscript

About a decade after Laddie's passing, his nephew Subroto Mukherjee, inspired by him, decided to forgo medical studies and joined RAF Cranwell to serve in the Indian Air Force (IAF) in 1930. Mukherjee later became the first Chief of the Air Staff of the IAF in independent India.[49]

Remembering Laddie Roy, India's Flying Ace in WWI

In December 1998, to mark the 100th anniversary of Laddie's birth, the Indian Postal Service issued a commemorative stamp in his honour. In 2019, to honour the contributions of Indians in WWI, the Government of India issued a series of stamps including those of air warriors Hardit Singh Malik, Errol Suvo Chunder Sen, Shirikrishna Chanda Welinkar and Indra Lal 'Laddie' Roy.

2

Message in a Battle

Gobind Singh, VC – the Hero of Cambrai

When WWI started in 1914, Indian regiments were enlisted by the British government as part of the British Expeditionary Force (BEF) to serve on the frontlines. Surprises awaited them at every turn. Thousands of Indian soldiers journeyed to Europe; most of them had never set foot outside of India. The soldiers travelled for survival, food and the lure of the little money they made, which they sent home to their families. On arriving in France, they were given a warm reception, with people cheering and embracing them. However, they encountered a Western culture that presented a stark contrast to the social norms they were accustomed to back home.

At the warfront for the first time, all they could see were drains, swamps and flat marshland. There was no enemy in sight; they remained concealed within their trenches a few hundred yards away from those of the British. The Indian soldiers had never engaged in trench warfare. Having fought in India's northwest frontier in the past decade, they were accustomed to employing fire and movement tactics and using the terrain for cover. Now, they were being told that the ground between them and the enemy would stay unutilized. Instead, the ditches and swamps would become the trenches from where they would wage the big war. Over the long winter in 1914, Indian soldiers ended up manning one-third of the British line in France.[1]

After the initial phase of trench warfare, a shift occurred as armies abandoned their positions and advanced towards the enemy. The infantry and cavalry clashed amid artillery shelling, and the result was a bloody stalemate.

As commander-in-chief of the BEF, General Douglas Haig had witnessed first-hand the futile battles on the flatlands of Belgium, such as Somme, Verdun, Poelcapelle and Passchendaele. Soldiers on opposing sides fought in blinding rain over muddy lands devoid of cover. These battles resulted in countless deaths but yielded minimal gains, and the Allies struggled to achieve a breakthrough.

By early 1917, a deadlock had arisen between the two sides. The Germans, firmly entrenched on French soil, withdrew to a new fortified position known as the Hindenburg Line (Siegfriedstellung in German). This line, which stretched 120 kilometres across north-eastern France, from Lens to beyond Verdun, was regarded as the German army's most formidable line of defence, and was believed to be impenetrable at the time.

The British launched several campaigns to eventually gain control of the line, including a plan to capture the German-held town of Cambrai, an important railhead in northern France. The operational plan was to deploy various elements of a battle. Cavalry, air power, artillery and infantry were strategically combined for the first time in history at Cambrai, and this approach was given the name 'all-arms warfare'.[2] Alongside these elements, the Allied Forces had devised a powerful industrial invention to be unleashed at Cambrai, aiming to overpower the enemy and win the war.

The Battle of Cambrai – which took place between 20 November and 7 December 1917 – changed the face of warfare and forms the backdrop for this story. However, despite the use of heavy weaponry, it was the bravery and courage of an Indian soldier that made a profound impact in this battle.

This is the story of how Lance Daffadar[3] Gobind Singh escaped death not once, not twice, but three times, crossing enemy lines to play a crucial role in saving countless lives.

Gobind Singh goes to war

In September 1917, Gobind Singh was in Persia with his regiment when move orders[4] arrived for him to be a part of the war in Europe.

Like the other young boys in Damoi, a tiny hamlet in Nagore district of Rajasthan, Gobind had grown up dreaming of wearing a uniform and going to war. His first opportunity to don a military uniform came in 1903, when he was recruited by Jodhpur Lancers, a cavalry unit of the erstwhile princely state of Jodhpur. The entire village of Damoi rejoiced. Standing at 5 feet 8 inches tall and adorning a bright safa (turban) atop his head, Gobind sported a strong, stocky physique and quickly acquired a reputation for being a calm and courageous young man.

The Jodhpur Lancers was among the notable polo-playing units during that time, and Gobind took a shine to the sport.

A nippy rider and crafty player on the field, he soon transferred to the 28 Light Cavalry regiment. When war broke out in Europe, the unit was mobilized from Quetta, a military base and training centre for British and Indian forces, to Persia, where they were posted till 1918.

From the modest village of Damoi to the opulence of Persia, Gobind, a Rathore of the Suryavanshi Rajput lineage, had travelled a long way. The young men and women in his village now regarded him with considerable awe and a great deal of izzat (respect).

Lance Daffadar Gobind Singh

When the call to join the war in Europe finally came, only a select few from 28 Light Cavalry were to be sent. Two officers, fifty soldiers (other ranks, i.e., who were not officers) and a hundred horses. When the list was drawn up, Gobind found his name on it. He had been selected to carry the Regimental Colours into battle, a rare honour.

Like Gobind and the soldiers from 28 Light Cavalry, other fresh arrivals at Cambrai enabled field commanders to infuse energy and enthusiasm into a dreary war-ravaged force. Alongside the reinforcements, General Haig used a strategic element to inject confidence and strength into their battle plans.

The Allied Forces discover the X factor

Leaning forward in his wickerwork chair, General Haig marvelled at the lazy gallop of agile horses of the cavalry before him.[5] He was a cavalryman at heart, whose faith in God and horses surpassed his opinion of other means of warfare, which he believed held no genuine value for a true warrior.[6] 'Bullets have little stopping power against the horse,' he was known to have said.[7]

However, the General had observed the devastating toll modern weaponry took, resulting in horrific casualties on all sides. A year into WWI, he grappled with a pressing question: How could the German defences be breached?

In 1915, two engineers embarked on an experiment to find an answer.

For a few years, troops had transformed vehicles into armoured cars by equipping them with guns. But these vehicles, with standard wheels, were limited to road travel;

they would get immobilized in swamps and ditches, making them sitting ducks for enemy infantry.

William Tritton and Lieutenant Walter Wilson were determined to address this formidable challenge. In a little-known White Hart hotel in Lincoln near London, away from the glare of the print media, the duo bolted themselves inside their room, debated designs and drew sketches on envelopes, cigarette packets and pieces of paper. The designs were rushed to the factory, and the production team worked diligently, knowing that there might be a race with the Germans to create a powerful machine that could change the course of the war.

The engineers kept the prototype vehicles parked behind the hotel, covered from prying onlookers. Their resemblance to steel water tanks helped them keep the secret.

These tanks, first deployed during the Battle of Flers-Courcelette, a significant phase of the First Battle of the Somme that occurred from 15 to 22 September 1916, performed poorly on boggy grounds. Some broke down, others got stuck in the difficult terrain or were disabled by enemy fire. By 1917, however, significant advancements in tank design and deployment strategies had already occurred. With the monsoons over and winter taking hold in Belgium, the Cambrai terrain offered an ideal chance to put these innovations to the test. The upcoming Battle of Cambrai in 1917 would witness the largest assembly of tanks to date.[8]

Despite their formidable presence on the battlefield, these steel behemoths couldn't shake the unsuitable alias they had acquired at the outset. Tanks – that's how they came to be known forever.

Unleashing the assault at Cambrai

Cambrai lay at a junction of railways connecting the towns of Douai, Valenciennes and Saint-Quentin in northern France. Situated near the Saint-Quentin canal, it served as a vital supply route for enemy forces, bringing in supplies from Germany and other parts of occupied France. General Haig and the British commanders knew that the capture of Cambrai would disrupt the German supply and communication network. There was, however, one snag – Cambrai lay behind a formidable defensive position of the Hindenburg Line. To break through that defence line, a strong offensive operation was required, followed by an infantry advance to seize the industrial town.

Battered by the Battle of Passchendaele, which had taken place between July and November 1917, and in which both sides suffered heavy casualties and endured appalling conditions, the Oberste Heeresleitung (German High Command) was intent on providing their troops with much-needed rest and recuperation. The Cambrai sector, with its deep trenches, dugouts and thick barbed wire of the Hindenburg Line, served as a 'sanatorium of the West'[9] for the exhausted German forces.

The British forces had established their positions and assembly areas in the regions surrounding Cambrai, preparing for the offensive. The peace prevalent in the sector allowed them to construct substantial trench positions and prepare the troops. Lieutenant Kenneth Page of 40 Brigade and 2nd Division remembered from that time, 'the Germans didn't shoot at us, and we didn't shoot at them'.[10]

The tranquillity of the 'sanatorium' was about to be blown apart.

Separately, Lieutenant Colonel J.F.C. Fuller of the Tank Corps and Brigadier H.H. Owen Tudor of the 9th (Scottish) Division each devised their own plans for an attack supported by tanks to cut through the German barbed wire and break the trench deadlock. Both plans were presented to General Julian Byng, the commander of the 3rd Army, a formation comprising British and Commonwealth troops established in 1916, and responsible for operations on the ground.

Meanwhile, General Haig, as the commander-in-chief of the BEF, held overall responsibility for strategic decisions and lent direction to the battle. General Haig and Byng worked together closely to ensure the success of the Cambrai offensive.

Tudor understood that by avoiding the loud artillery bombardment at the start, an attack could surprise the enemy and catch them off guard. Tudor's plan focused on a quieter approach to pinpointing and engaging targets instead of the typical noisy range-finder shots. Impressed with the ideas, Byng drew up a detailed plan of attack.

Nineteen infantry divisions along with five cavalry divisions were assembled, along with a secret deployment of the newly developed Mark IV[11] tanks – an improved version of the tanks previously employed at Flers-Courcelette with better armour protection, improved mobility and enhanced firepower. In total, the British deployed 476 tanks at Cambrai, including 378 in combat roles.[12] Four Royal Flying Corps (RFC) Squadrons provided air support.

The opening skirmish of tanks and infantry

Before 19 November 1917, the British frontline in the Épehy sector, a strategic battleground, extended northwest of St

Quentin, bent around to the southwest of Gonnelieu and Villers-Guislain and then bypassed Little Priel Farm, Le Tombois Farm and Gillemont Farm. The forces were familiar with the area: in the previous year, two men had won the Victoria Cross for bravery in the sector[13] (see map on page 43).

On a dull, overcast morning of 20 November 1917, a deafening roar of hundreds of tanks broke the tense silence. Rolling down the slopes, the tanks motored effortlessly along a 10 kilometres front, charging at the Hindenburg Line while the accompanying artillery bombardment softened up the German defences. Second Lieutenant Edward Leigh-Jones from the Tank Corps vividly remembered the daunting challenge posed by the formidable Hindenburg Line. 'It was 50–100 yards deep, the wire, and we anticipated that it might be a serious obstacle.'[14] Tanks, however, took apart the obstacles and wire fences, and split open German defences, crashing past them. The thunderous sounds and terrifying sight of hundreds of oval-shaped Mark IV tanks caused shock and awe among German defences. Within hours, the British forces took eight thousand prisoners and a hundred guns, forcing the German army back on the first day of the offensive.[15] By the evening, the British forces were closing in on Cambrai.

Victory seemed around the corner.

For the plan to be successful, however, it required infantry forces to advance rapidly, build on the breakthrough created by the tank onslaught and capture enemy ground.

While tanks managed to breach German defences, they couldn't hold the captured territory due to delayed arrival of infantry forces.

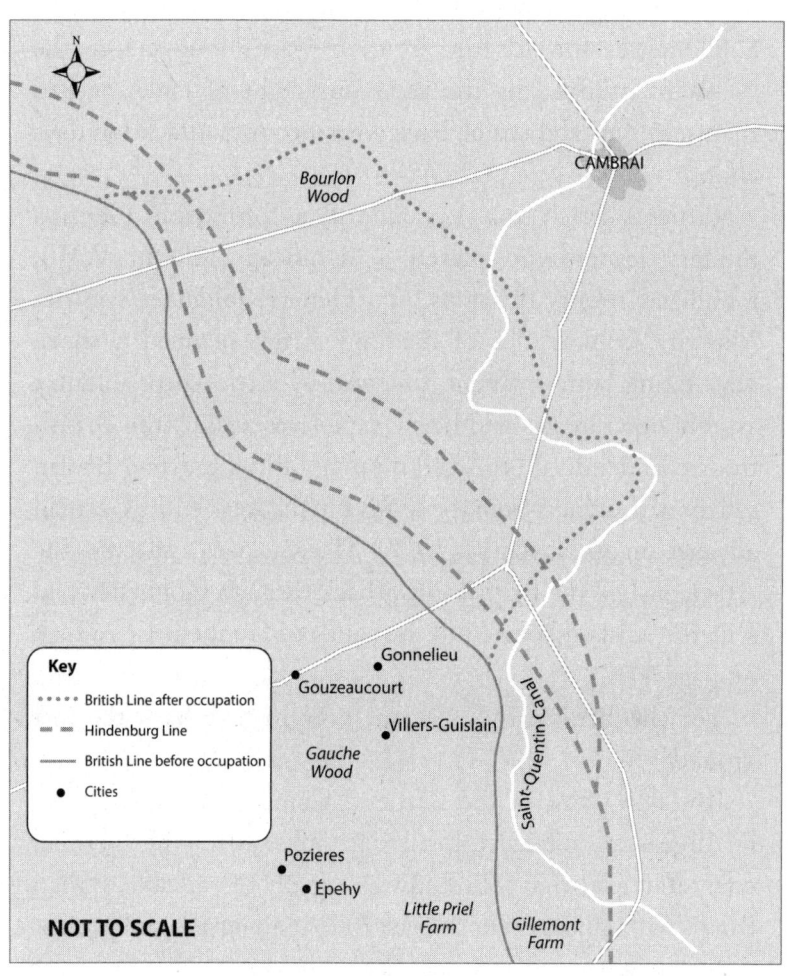

Key Locations in Battle of Cambrai, 1917

The movement of troops, equipment and supplies, and congestion on the roads had caused a sluggish arrival of reinforcements. Troops took 15 hours to cover the final 5 kilometres towards the front. It wasn't long before the Germans, stunned by the early onslaught of tanks, picked themselves up and struck back with a counterattack ten days later.

General Erich von Ludendorff, a formidable German military leader who played a significant role in WWI, mobilized twenty divisions, and General Johannes von der Marwitz, commander of the 2nd Army, planned a short, devastating bombardment, followed by a two-corps infantry assault aided by gas and heavy explosives against the British troops. Ludendorff proposed nipping off the growing British salient with attacks from the south and the west.[16] The German advance on 30 November, led by Marwitz, was a success. The attack pushed the British forces back through Gonnelieu and Villers-Guislain, located to the south of Gonnelieu (see map on page 51).

For the British, the aim was now to take back the lost ground.

The first phase of the Battle of Cambrai was notable for its large-scale offensive involving tanks. After the German counterattack, what followed was equally remarkable. Indian troops with lances on horseback fought alongside and against the most advanced technological innovations in warfare at that time.

Lance Daffadar Gobind Singh was about to be called into action.

Charge of the Indians

From 28 Light Cavalry, Gobind had been attached to 2nd Lancers (also known as Gardner's Horse), a renowned cavalry regiment primarily composed of Sikh soldiers. The 2nd Lancers – along with other cavalry unit 6th (Inniskilling) Dragoons and 38th Central India Horse – went to France as part of the 5th (Mhow) Cavalry Brigade, in the 1st Indian Cavalry Division.[17]

The Germans had reoccupied the territories of Gouzeaucourt, Gauche Wood and Villers-Guislain, located south of Cambrai. The task of the cavalry was to retake these territories. On 30 November, 5th Mhow Brigade and 4th Cavalry Division, another Indian cavalry regiment deployed in Cambrai, were instructed to launch a mounted attack to recapture Villers-Guislain Ridge.

The Brigade Headquarter (HQ) was located at Épehy and Pozières, under the command of Brigadier Neil Haig (cousin of General Haig). Villers-Guislain, Épehy and Pozières, all part of the Cambrai region, form a triangular relationship in terms of their positions on the map. Villers-Guislain is to the north, Épehy is east and Pozières is west.

Brigadier Haig and Major General Alfred Kennedy, the 4th Cavalry Division commander, had observed enemy machine guns in action at a strategic location southwest of Villers-Guislain. They realized that any attack would face heavy fire and be easily suppressed. However, the Cavalry Corps commander, Lieutenant General Charles Kavanagh, had a limited understanding of the prevailing ground conditions and did not possess the patience of a seasoned warrior. He ordered Brigadier Haig's brigade to attack the Germans. The three battalions in the brigade, accompanied by a section of the 11th Machine Gun Squadron, advanced closer to the German trenches.

The attack was planned with a two-pronged approach. The main attack was to be carried out by the 6th (Inniskilling) Dragoons and a subsidiary attack by the 2nd Lancers.[18]

The Inniskillings pressed forward, but heavy machine gun fire from a sugar beet factory, southwest of Villers-Guislain, forced them to retreat to Pozières with significant casualties.[19]

On 1 December, at 0800 hours, 2nd Lancers received orders to proceed. Their lances arrived just in time from the rear, lugged forward by a cart. Armed, the men hurriedly lined up for the attack.[20] Captain Dysart Whitworth of 2nd Lancers led the columns, accompanied by a squadron from 6th (Inniskilling) Dragoons and a mounted two-gun section of 11th Machine Gun Squadron, from Pozières through Épehy.

A hundred yards on, the Lancers crossed the forward trenches – the first line of defence – and organized themselves into open columns of squadrons.[21] Leading the advance was the C Squadron commanded by Major Knowles.[22]

While the main objective was capturing the formidable Villers-Guislain Ridge, this unit had a crucial task ahead. They needed to secure the strategic high ground at Targelle Ravine, east of Villers-Guislain, by advancing through Kildare Lane (Trench), one of the German trenches. Lieutenant Colonel Turner, commander of the 2nd Lancers, and Knowles took different routes for their advance.

An hour and a half later, at 0930 hours, they unleashed their attack.

Approximately 1.5 kilometre ahead of the headquarters, there was a cluster of four elm trees known as Limerick Post, which had been captured by the Germans the previous day. As the squadrons charged through the shallow valley covered in shelling, they were met with gunfire from German machine guns positioned in Kildare trench, as well as from enemy outposts on their sides. Amid the rat-tat-tat of machine guns firing from the enemy trenches, the horses raced through the expansive valley.

The sight of Gobind and the audacious, galloping Indian cavalry soldiers with lances had unnerved the machine gun–wielding Germans. 'The riders are coming quickly closer, under heavy fire from trench positions to left and right. But they are absolutely fearless and are put off by nothing. Now, they are coming right out of the depression. And now I also know who I have before me. Indians, real Indians, savages, the brown devils from newspaper reports,' shouted Lieutenant Karl Christian of the German Infanterie-Regiment 418, which held the occupied trenches.[23] Upon witnessing the cavalry charge, some machine gunners hastily abandoned their trenches in fear. 'Shoot! Get the guns on to them,' Christian cried as he rallied his troops against the Indians.[24]

Jemadar Jiwan Singh of 2nd Lancers had been thrilled to be in action for the first time in three years. (Jemadar is similar to a subedar.) He vividly recalled the events of that day. It was an unusual sight: intrepid Indian cavalry soldiers with lances charged ferociously at the nervous enemy in the trenches that kept firing away with their machine guns. Raw courage had been pitted against cold steel. 'The fury of our charge and the ardour of our war cries so alarmed the enemy that he left his trenches and fled. At first, we were assailed by machine guns like a rainstorm, but how could the cowardly Germans stand before the onslaught of the braves of the Khalsa!'[25]

As the horses advanced, they encountered stretched barbed wires obstructing their path across Kildare trench, requiring the men on horseback to courageously navigate through three lines of 18-inch-high barbed wire gaps.[26] The wires disrupted their forward momentum. The desperate German machine gunners, who had thus far been unable to hit the brigade's nimble horses, had eagerly awaited this opportunity. They unleashed a series of bursts at the leading columns, resulting in Turner and his staff bearing the brunt of the gunfire. Turner was killed instantly. The 2nd Lancers had no time to mourn the demise of their courageous and beloved leader, affectionately known as 'Lion-hearted' or 'Sherdil' to his Indian troops.

Indian soldiers, including Gobind and Sowar Jot Ram, dismounted and valiantly defended the position they had taken. (Sowar was the rank for Indian cavalry troopers or privates.) The area became a tangle of men and horses. Amid the chaos and enemy fire, casualties mounted, but the men fought on with determination.

There were about 50–100 Germans occupying the trench. Fifteen of them were killed by sword or lance and 20–30 by rifle or machine gun fire. The Germans were in disarray.[27] The captured position was eventually held by 200 men of the 2nd Lancers and 36 men of Inniskilling.[28] Isolated from the rest of the brigade and facing a swift German counterattack, however, they faced a dire situation.

Round 1: An urgent message needs to be sent

At the Brigade HQ in Pozières, concern grew that the 2nd Lancers might have been surrounded and overwhelmed by the German forces. In a bid to provide support, three dismounted squadrons of the 38th Central India Horse were dispatched. However, they encountered heavy enemy fire and struggled to make significant progress.

On the other hand, time was running out for the 2nd Lancers. Knowles, now in command following Turner's death, realized the urgency of sending a message to Brigadier Haig at Brigade HQ, informing him of the dire situation at Kildare trench. The only means of communication was through a runner, as the Germans had regained their positions and were ready with their machine guns, leaving no room for the beleaguered Lancers to breathe. Without timely intervention, they faced the grim prospect of being overwhelmed by the superior German forces.

Knowles, stationed in the trench, called for volunteers. 'Who will go?'[29] he asked.

Two men immediately stepped forward to answer the call. Sowar Jot Ram and Lance Daffadar Gobind Singh.[30] The two had been given identical messages to deliver.

The distance from their location to Pozières was 2.5 kilometres. The route was treacherous, requiring them to cross open ground on horseback. They charged forward at a gallop, each taking a different path.

Jot Ram took the shorter route. As he passed the Limerick Post, the machine guns opened fire, killing him instantly. Gobind, on the other hand, darted down to the lower ground to the south, sweeping past the hail of bullets. After covering half a mile, his horse was hit by a burst and they both fell to the ground. The brave animal died, but Gobind had survived. Possessing a warrior's instincts and unwavering composure, he lay dogo (military term for laying quiet), feigned death on the ground, and waited for the danger to subside. Amid the heavy gunfire, a brief silence fell. Seizing the moment, Gobind cautiously rolled across the ground, gaining precious distance. With calculated timing, he rose and sprinted, evading enemy fire.

Displaying unshakeable composure, Gobind repeatedly changed the course of his run, dropped to the ground feigning death before getting up to sprint a few yards in a zigzag manner, and then hit the ground again to deceive the enemy. In the act of dropping and rolling on the ground, he was creating a smaller, more difficult target for them. His subtle movements and cunning tactics frustrated his opponents, as their shots missed their mark. With every yard he advanced, he stealthily closed in on the HQ location. Finally, to the astonishment and joy of his comrades, the resilient Indian soldier reached the headquarters.

Gobind swiftly reported the unfolding events. The Brigade HQ had the critical task of sending instructions to the 2nd Lancers' location and addressing their previous message. It was a tenuous situation as the enemy had targeted messengers along the route. Sowar Jot Ram had been killed, and Gobind had barely managed to evade death and survive the dangerous journey. Now, the enemy had their guns aimed at any messenger travelling the route. The element of surprise was gone. The superiors at Brigade HQ faced a difficult choice (see map on page 51).

Round 2: Brigade sends Gobind Singh to deliver a reply

Gobind was handed the message and a fresh horse to make his way back. It marked the beginning of another perilous

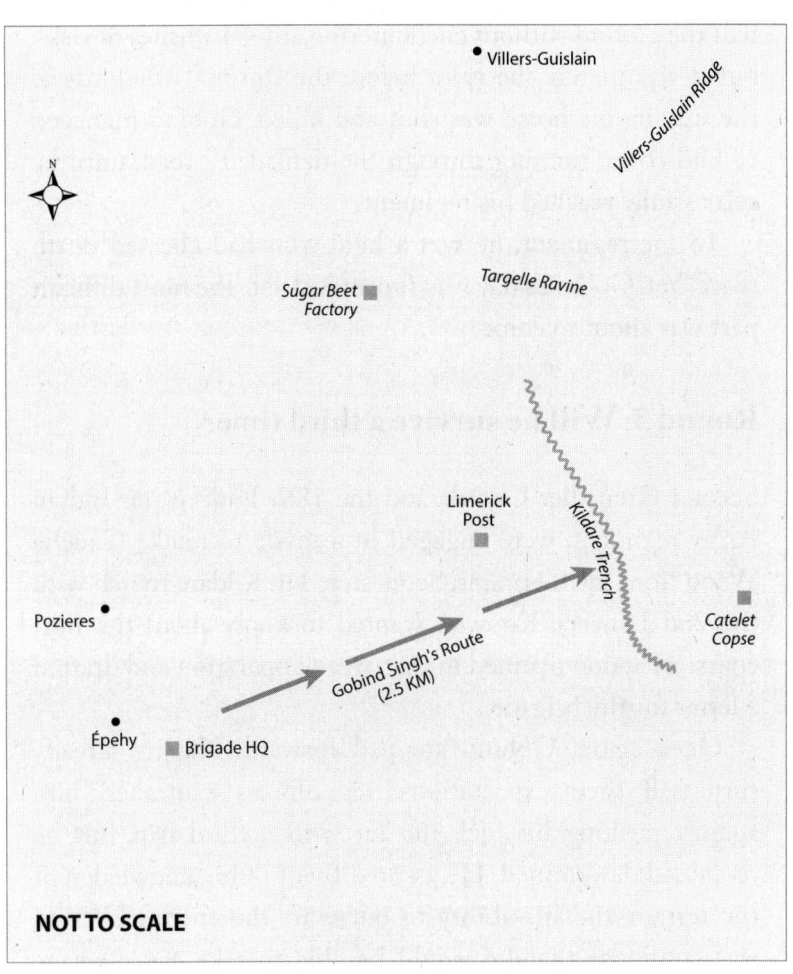

Route Taken by Gobind Singh to Carry the Message

journey. This time, he chose a higher ground south of the valley, skilfully navigating and avoiding direct enemy fire until he was close to the German post. He had covered more than half the ground without encountering any skirmishes or risks, but it was merely the calm before the storm. Two-thirds of the way in, his horse was shot and killed. Gobind managed to find cover, running through the defiladed[31] road, until he successfully reached his regiment.

To the regiment, he was a hero who had cheated death twice. Yet, Gobind's day wasn't over just yet. The most difficult part was about to come.

Round 3: Will he survive a third time?

Second Grenadier Guards and the 18th Lancers, an Indian Army regiment, were engaged in a tussle to retake Gauche Wood from the Germans. Sequestered in Kildare trench with the 2nd Lancers, Knowles wanted to know about the next course of action planned in that area of operation and drafted a letter for the brigade.

Once again, Gobind stepped forward. Having already surpassed their expectations, his officers cautioned him against pushing his luck too far with a third trip, but he remained determined. He assured them of his knowledge of the terrain and his ability to outsmart the enemy. Gobind was confident that he would be able to take the message successfully to the HQ.

The officers relented, allowing him to select a horse of his choice. Those present feared that this would be his final act, his 'swansong'. They bid Gobind a heartfelt farewell and wished him luck.

Gobind's flying figure galloped into the mist as he chose a different route this time, guided by the direction of Catelet Copse. He encountered an initial barrage of well-directed enemy fire. This time, the Germans were ready for him: Gobind was surely riding towards his death. Halfway through the journey, a sudden explosion shook the air and artillery shells fell around him. One of the shells struck his horse and shattered the animal. From the trenches, his fellow soldiers watched in horror as Gobind and his horse disappeared into the thick cloud of smoke and dust created by the intense barrage. His third horse had died under the onslaught of heavy firing; his regimental soldiers believed Gobind had perished as well. That night, he was declared 'Missing in Action'.

Miraculously, Gobind had survived the ordeal. Falling hard on the ground, he staggered onward, tired and dazed. Despite exhaustion and injuries, he defied the determined enemy machine guns that sought to bring him down, weaving and twisting to evade the relentless firing around him. This time, the guns had shown a clear determination to eliminate him. As the Germans unleashed a barrage of shells that struck his horse, they believed they had successfully taken him down as well.

Just before noon, the sentry at Brigade HQ saw a lone, blood-spattered figure lurch into view, carrying a piece of paper sprayed in crimson. The soldiers who received him thought Gobind had been severely injured and were astonished to discover that the blood belonged to his fallen horse, who had died inches away from him. It was a miracle that he had survived.

Shortly after, a response to the message from Knowles needed to be sent to the 2nd Lancers. Despite his exhaustion and injuries, Gobind put his hand up again. His enthusiasm left the officers speechless! Brigade HQ would have none of it. Gobind was a legend now, and they weren't ready to lose one so quickly. They turned him down.

Victoria Cross for the valiant messenger

The fighting in the Épehy–Villers–Guislain region reached a stalemate as neither side was able to achieve a breakthrough. With no further advance in sight, the brigade troops were withdrawn.[32] The following day, the 2nd Lancers moved south to the outskirts of Saint-Émélie, a village 7.5 kilometres from Épehy, and thereafter on 3 December 1917, the Mhow Brigade marched back to the billets (temporary accommodation) to rest and reorganize.

By 5 December, the British had been driven back almost to their original positions. After two weeks of bitter fighting and casualties of about 45,000 each for the two sides, the Battle of Cambrai reached a stalemate.[33] Although the British didn't fully capitalize on the early achievements of their tanks, the battle highlighted the crucial role of armoured vehicles in determining outcomes on the Western Front.

For his outstanding and unmatched gallantry, Lance Daffadar Gobind Singh was awarded the Victoria Cross, the highest military decoration for valour in the face of the enemy. On 6 February 1918, during the investiture ceremony at Buckingham Palace, he received the decoration from the King himself. He was also honoured with a reception, where he was joined by two Indian cavalry officers who were visiting London as guests of the nation. As part of the recognition for his extraordinary bravery, he received a piece of silver plate and a gold watch. The reception was attended by General Sir Garrett O'Moore Creagh, a decorated veteran and former commander-in-chief of India, and Lieutenant General Sir Pertab Singhji, a highly esteemed soldier with extensive wartime experience, including several campaigns in WWI.

The Victoria Cross awarded to Gobind Singh

His citation, published in the *London Gazette*, read: 'Gobind Singh, Lance Daffadar, Indian Cavalry. For most conspicuous bravery and devotion to duty in thrice volunteering to carry messages between the regiment and Brigade Headquarters, a distance of one-and-a-half miles over open ground, which was under observation and heavy fire of the enemy. He succeeded each time in delivering his message, although on each occasion his horse was shot and he was compelled to finish his journey on foot.'

Gobind's audacious acts and the bravery of the 2nd Lancers and the Indian cavalrymen had thwarted a major enemy breakthrough, saving the villages of Épehy and Pozières from continuous shelling and preventing an enemy advance. The brave lancer's consecutive runs delivering timely messages saved countless lives and ensured the regiment's survival.

Today, in Damoi, he is still revered as 'VC Sahab' (Victoria Cross Sahab), and his twelve-room haveli (home) stands as a proud testament to his heroism. There are two large frames with his painted portraits in army fatigues in the courtyard. His fierce, proud eyes fix visitors with an intent, clear gaze, serving as a powerful reminder of the messages he repeatedly risked his life to convey over a century ago.[34]

Postscript

During the war, 1.5 million Indian soldiers journeyed from remote villages in India to the battlefields, surpassing the combined armies of Scotland, Northern Ireland and Wales.[35] They left behind a proud legacy of courage and commitment. Several Indian cavalry regiments fought in the Battle of Cambrai, and each year, they observe Cambrai Day on 30 November and 1 December. Lance Daffadar Gobind Singh, hero of Cambrai, was among the remarkable soldiers who demonstrated extraordinary bravery during WWI.

Interestingly, only a few weeks before Gobind's courageous actions, a similar incident took place against the Hindenburg

Line. Alfred Mendes, who served in the 1st Battalion Rifle Brigade and fought in the Battle of Passchendaele in 1917, had a unique connection to courage as well. At Passchendaele, he heard the extraordinary story of two messengers who travelled across miles of battleground to deliver an urgent message that saved 1,600 of their fellow soldiers from a massacre.

Many years later, Alfred shared the amazing story he had heard in the war with his grandson, film director Sam Mendes. In 2019, nearly a century after the episode, Sam directed a Hollywood movie on that story titled *1917*. It earned ten nominations and three awards at the Oscars, further immortalizing the tale.

3

Three Lives in a War

The Miraculous Escapes of Chanan Singh Dhillon

On a cool summer morning in June 2018, a local train bound for Limburg an der Lahn in Hesse, Germany, departed from Frankfurt station. As the train picked up pace – gently at first – before hurtling through the outskirts of the city, and the nearby hamlets squeezed into a blur outside the window, Gurbinder Dhillon's thoughts wandered towards the purpose of this journey. After several unsuccessful attempts to find answers – contacting a military attaché at the Indian embassy, German officials and bureaucrats – an unexpected saviour had emerged.

Vladimir Magyar, a former Czech army veteran and friend, had heard about Gurbinder's search. One day, he arrived with a map in hand, laid it open and pointed to a place on it. 'What is this?' Gurbinder asked. 'That's the place you are looking for – in Limburg. Stalag 12A,' Magyar said, smiling.

Indeed, Gurbinder's journey, destined to culminate in Limburg, had been set in motion by a story in his father Chanan Singh Dhillon's World War II (WWII) diary.

Eight decades earlier, in 1939 . . .

It was a crushing rejection.

At the Services Selection Board, HQ Western Command, Rawalpindi, a twenty-one-year-old Chanan appeared before a British officer with eager anticipation. He yearned to fulfil his childhood dream of becoming an officer in the British Indian Army, holding a leadership position and commanding troops. He had fared well in leadership tests at the selection but performed poorly in the discussion phase.[1] To his disappointment, he was told that his English proficiency fell short of the standards required to qualify as an officer.

The realization stung deeply, but Chanan found solace in the bitter truth that the army he aspired to serve wasn't truly India's, as the British held dominion. Growing up in Punjab, his father, a farmer, often worried about how his son would manage the struggles of farm life. The army provided a ready escape from those challenges, offering employment, income and a safety net.

Despite the rejection, Chanan remained determined to pursue his dream of becoming an officer. Enlisting as a soldier in the unit, he underwent rigorous combatant and fieldwork training, determined to prove his mettle. He had also decided to embark on a quest to master the English language, knowing that it held the key to unlocking his path to leadership. In his diary, Chanan wrote, 'Reading Urdu novels was my pastime; I switched over my interest to English books. This was not easy. The changeover was so painful and slow that at times I was disheartened and did not know what to do. However, I persevered and registered steady progress. In addition, the President of the Selection Board gave me a letter for my officer in charge in which he requested him to make sure I was up to date with my spoken English so that I could reappear.' Captain Radcliff Smith, the officer in charge, even arranged for an English tutor.

Little did Chanan know that a development unfolding in a faraway place was about to alter his destiny.

In the fall of that year, on 1 September 1939, German dictator Adolf Hitler launched an invasion of Poland. Over 1.5 million German soldiers, accompanied by over 2,000 airplanes and more than 2,500 tanks, marched into the country. In response, Great Britain declared war on Germany, marking the onset of a global conflict – WWII.

'As the unit was being mobilized, there was a change in command and Captain Radcliff Smith was transferred on promotion. Before moving on to the new posting he recommended me to his successor to kindly see that I was again recommended for commission as soon as the unit settled down,' Chanan noted in his diary.

Chanan's unit – 41st Field Park Coy, RE, 8th Army – embarked on a nine-day journey by ship from Bombay (now Mumbai) to Basra in Iraq. Their mission was to secure the Persian Corridor, an important supply route for the Allies. En route from Bombay to Basra, Chanan was promoted to Naik (a non-commissioned officer rank equivalent to corporal) and picked up his stripes. The CO also announced that all English-speaking ranks would wear badges on their shoulders, with the alphabet 'E' embroidered on a red background. Standing on the deck with the badge and the stripes adorning his uniform, Chanan smiled to himself as the ship sailed through the high seas. He was rising fast in the army.

Naik Chanan Singh Dhillion in WWII

While travelling with the regiment through Iran, however, doubts began to creep in about the true purpose of the war he was involved in. This conversation was recorded in his diary:

'How wonderful it would be if we rushed by road through Baluchistan to join in the freedom struggle for our motherland,' whispered Chanan to himself, overcome with emotion about the movement back home that was gaining momentum.

'What were you muttering about Baluchistan . . .?' asked a fellow soldier.

'Nothing . . . I was wondering if Baluchistan is nearby . . .?'

'Yes, it is . . . so close we could be fighting the same battle across two countries . . .' The soldier winked, alluding to the irony of fighting against the British in a foreign land while also serving alongside them.

'I see . . .' Chanan nodded.

Deep down, he wanted to fight the colonial power back home. Instead, he had signed up to defend the Union Jack in the war. He learnt to keep his thoughts to himself. Soon, Chanan and his unit found themselves in Africa, a relatively quieter place compared to the raging conflict in Europe.

A letter arrives

In 1940, one year after declaring war on Germany, Great Britain was facing significant challenges in WWII. Their ground forces in Europe were being overrun by the advancing German army. However, the newly elected Prime Minister Winston Churchill firmly believed that the British Army needed one victory to turn the tide in their favour. That opportunity presented itself in North Africa.

Determined not to be overshadowed by his friend and ally Adolf Hitler, Italian dictator Benito Mussolini invaded and occupied Abyssinia and Libya as part of his expansionist ambitions and to establish Italian dominance in North Africa.

Fuelled by his success, Mussolini had set his sights on Egypt as his next conquest.

With 0.5 million Italian troops poised to attack, the odds had been stacked against a British force of 36,000 men defending northern Egypt,[2] an independent state under British control. The British were faced with the daunting task of defending Egypt's strategic position, including the vital Suez Canal, against the advancing Italian forces.

Italian troops broke through British defences but instead of capitalizing on the advantage, they inexplicably halted 60 miles inside Egypt, creating a thin front and leaving significant gaps that the British attacking forces exploited. Due to these tactical errors made by the Italian forces,[3] the British launched a counteroffensive that allowed them to recapture territories in North Africa, including Libya.[4] Churchill's words rang true: the army had delivered a turnaround.

Unbeknownst to them, however, a more formidable foe had followed the events closely, and the British were about to face a tougher challenge.

In a swift response to Britain's expanding foothold in North Africa, Hitler acted. In February 1941, he sent the Afrika Korps, led by his ace commander General Erwin Rommel, to counter the British. By January 1942, Rommel's forces had pushed the British back to the Egyptian frontier and captured Benghāzī in Libya. The German and Italian troops planned to attack British defences at El Alamein in Egypt by 30 June 1942.[5] Africa was becoming the backdrop for a big fight.

Amid this raging war, on 20 June 1942, a letter arrived addressed to Chanan Singh Dhillon, whose unit was posted in Mersa Matruh, a coastal township situated near El Alamein in North Africa. Lieutenant P. Devonald, his platoon commander, handed it to him. 'I have good news for you,' said the British officer, with a toothy grin. The letter had orders for Chanan to sail to India to join the training academy.

Chanan's eyes filled with tears; he had never let go of his dream and finally, his moment had arrived. He was about to fulfil his aspiration of becoming a commissioned officer. Filled with a sense of triumph and nervous anticipation, Chanan eagerly counted down the days until the ship that would take him home would arrive.

Just five more days to go.

On 28 June 1942, there occurred a significant turn of events.

As the battle shifted to North Africa, Chanan and his unit valiantly defended their position at Mersa Matruh. News arrived one evening that Rommel's German army was closing in.

On 26 June, they faced a formidable clash with Rommel's forces, engaging in fierce fighting that persisted for three days. With British defences stretched, Rommel's attacking columns steamrolled through the fortifications, moving with blazing speed.

At approximately 2350 hours on 28 June, Chanan's unit received the order to retreat. He and his mates planted dynamite to destroy the installations – such as bunkers, fortifications and other defensive structures – before their departure. They consulted the withdrawal plan on the map. They had to travel six miles on the main road towards Cairo, then turn right and head south into the desert.

The plan was to go approximately 32 kilometres deep into the desert, make a semicircle towards the East and join the road using a railway track. This would place them farther away from the advancing Germans.

Chanan had thought this would be his last piece of action before the ship to take him to India arrived. He would then return to the war after training and gaining a commission – as a lieutenant leading his men in battle.

From combat to captured-in-the-desert

Not content to simply hold the captured territory, Rommel's troops had pursued the retreating British soldiers. Determined to escape, Chanan and his comrades sped away in their vehicles, racing through the desert to evade the Germans.

Their detour into the desert had been a strategy to outmanoeuvre the Germans and outlast their pursuit. But the Germans had other ideas. As Chanan and his group drove on, the sound of gunfire grew nearer. As they abandoned their planned route, the lead vehicle in their convoy was ambushed and destroyed. The rest of the vehicles quickly changed direction, with Chanan's vehicle now leading the way.

They spent an uneasy night in the desert; by morning, they seemed to be in the clear, but continued to drive at a cautious and slow pace, remaining vigilant for any signs of danger.

At about 1030 hours, Chanan spotted two shadowy figures in the distance. As their vehicle approached, he realized they were two British armoured cars parked side by side, facing north. He turned to the driver beside him, letting out a deep sigh.

'Stop! Stop!' he said, raising his palm. The driver pressed the brakes but kept the engine running.

They stared in silence for what felt like an eternity. Were those abandoned or captured vehicles? Were there enemies hiding inside?

Chanan's gut said they were adversaries waiting for them. With nowhere to escape, all they could do was wait. The driver retrieved a bottle of water, a rarity in the desert. Chanan took a swig and felt the cool water ease his parched throat. He smiled nervously at the driver, aware that it might be a while before they had water again, if they survived.

The armoured cars remained motionless. Were they lying in wait, their gaze and weapons fixed on the convoy? Both sides held their positions, waiting to see who blinked first.

Chanan made the first move. 'Let's march,' he told the driver.

As they began to move, the armoured cars suddenly sprang to life and someone inside began firing. It was clear that these British vehicles had indeed been captured by the Germans.

Amid a fierce exchange of fire, the Germans swiftly closed in, targeting the rear vehicle in the British convoy just as it veered away. The vehicle, which had been hit, came to a screeching halt. Exiting, the men sprang into action, engaging the armoured cars with machine guns and light rifles.

Chanan's vehicle managed to escape, but the one behind it, carrying ammunition, came under German fire. The resulting explosion shook the desert, giving the scene the semblance of a minor battle. In retaliation, the British successfully incapacitated a German armoured car, forcing its crew to abandon it in the desert.

Though their transport had been damaged in the explosion, Chanan and six soldiers managed to escape by jumping into the last remaining vehicle and driving away. Two enemy jeeps raced towards them, but Chanan told the driver to press on, determined to make an escape.

Miraculously, an airstrip materialized, and the group saw an aircraft with British markings on the tarmac. Taking cover behind the sand dunes, they raced towards the airstrip.

Chanan understood it would be difficult to survive the exchange of fire, so he and his men quickly disembarked and set off on foot.

Suddenly, their path was obstructed by two German trucks brimming with soldiers. A stern-looking robust man leaped out and pointed his rifle at them.

'Down, down!' he shouted, and instructed them to kneel down. 'Now!' he said, in a menacing tone.

As they knelt with their arms raised, Chanan looked up at the soldier who held them at gunpoint.

It was all over.

They had fought bravely as a unit; surrendering was a bitter pill to swallow. Chanan's emotions surged like waves crashing against the shore, soaring briefly only to be shattered against the rocks in the next moment. 'Our fate is sealed, and the privileges granted to soldiers in active service will be denied to us. Pay, allowances, promotions, other benefits granted during the war, all gone . . . gone,' he lamented as he marched alongside the others captured. Initial shock was giving way to sadness.

He was a prisoner now.

Surviving captivity and challenging injustice

The POWs were at El Daba airfield, which had fallen into enemy hands a day earlier. As the long line of solemn captives marched forward, Chanan caught sight of the railway track featured in their retreat plan. He had originally meant to arrive at this location of his own volition. Instead, here he was, travelling in captivity.

Questions swirled in Chanan's mind: 'Does true freedom come from the power to choose?' Possibilities before him included: boarding the nearby goods wagons on the railway track, embarking on a ship back to India, or daring to escape. These thoughts both deflated and ignited his fighting spirit; the anger of losing his agency fuelled his determination. Amid it all, there was a realization that energized him. He could breathe. He felt alive.

A group of armed soldiers stood near the wagons. They held rifles and barked orders. The guards looked tired and unhappy like the prisoners but seemed 'free'.

Herded like sheep, the POWs trudged forward. Chanan's steps grew heavy and, eventually, his legs refused to carry him any further. The German soldiers cursed them loudly, their voices sounded more abrasive than the unfamiliar language. Chanan scanned the line-up of his men and did a mental head count. 'Seventeen from the company.'

He felt drained by the heat and his own exhaustion, causing the faces before him to blur together. Havildar Bhagwan Singh discreetly handed him a can of water, hidden beneath his shirt. After a furtive glance around, Chanan accepted it, feeling refreshed from the small relief it provided. Standing next to him, Devonald swayed, his usual cheer missing. He seemed lost, Chanan thought, wondering: 'Does one cease to be an officer once captured?'

Chanan had missed the opportunity to return home and

become an officer. Devonald expressed his remorse. 'Sorry . . .' he repeatedly said in a low voice. His eyes were moist.

※

After their capture by the Germans, Chanan and his comrades were handed over to the Italians, who took charge of the POWs and required labourers. The POWs were then transported to a camp located in Benghāzī.

Chanan's life had come to a standstill. He should have been in the training academy in India, but instead he was trapped in North Africa, unable to move forward and uncertain of what lay ahead.

Almost a month into their time at the camp, in the first week of July 1942, the morning alarm sounded earlier than usual. Several POWs were in the camp, but only Chanan and twenty-one other non-commissioned officers (NCOs) were bundled into a waiting lorry in the darkness and taken to a small fortified camp in Agela, a seaport. Chanan was assigned the routine task of loading ammunition on trucks as well as refuelling tankers from the diesel dump, a storage facility where large quantities of diesel fuel are stored.

The following morning at the diesel dump, the Italian soldiers barked orders at the POWs, instructing them to roll the diesel barrels out of the pits, transport them up a dirt ramp and load them into tankers for refuelling purposes. The German supply line had stretched from Tripoli (Libya) to El Alamein (Egypt) and the Italian guards at the camp were under pressure to ensure the delivery of supplies to support the German forces in the region.

The new arrivals had been overworked, and were experiencing cramps, pain and stiffened limbs resulting from dehydration. Chanan's shirt stank, as he hadn't had a change of clothes since the time of his capture, and his body was streaked with dirt and grime.

One afternoon, after a period of enduring monotonous routines and meagre meals of stale chapattis dipped in coffee, the exhausted POWs decided to stop work. 'This is against the rules of the Geneva Convention,' insisted Chanan, referring to the document signed by Germany in 1929. According to the convention, the work given to POWs should not be directly related to military operations.[6] Along with nine angry POWs, Chanan raised a protest. In the absence of a common language, they conveyed their dissent by raising their hands in the air.

An angry Italian NCO lunged at them and one of the sentries fired his rifle up in the air. The POWs panicked at the sound of fire. Aware of the toll the war had taken on everyone's nerves, Chanan took a step back, choosing to wait patiently for a more opportune moment.

The POWs resumed their work. That night, they were surprised when they were served macaroni for dinner, instead of the usual chapattis.

The next morning, they awoke to blisters the size of marbles. As the day progressed, the strain of work became too much for many of the POWs. They struggled to pick up the spades needed for scraping the dirt from the top of the barrels. By midday, their palms were coarse and bleeding from the laborious tasks. A couple of them even collapsed from exhaustion. The prisoners communicated their frustration through hand signals, as neither side understood the language of the other. In response to their protests, the angry NCO warned them of dire consequences.

On 20 July 1942, the POWs were hard at work at the diesel dump. 'They keep whipping us as shepherds do to their sheep,' is how Chanan describes it in his diary. At 1000 hours, the prisoners again stopped work. This time, the Italian NCO reported the mutiny to his officer, who demanded his Bren gun, a light machine gun. Chanan and his mates were ordered to line up.

Too tired to protest, they did as they were told. The officer made a last-ditch attempt to threaten them, signalling they would be shot if they didn't obey. The prisoners didn't care any more; so many had died already.

'Sooner or later, one has to die,' Chanan thought. Lips pursed, fists clenched and eyes tightly shut, he no longer felt the blisters. 'Put me out of misery,' he whispered loudly.

As the officer aimed at the line of human targets, Chanan closed his eyes. 'So this is how it all ends ... 1 ... 2 ... 3 ... !'

He waited. Would the echo of his own demise reach his ears? Seconds passed, and all he could hear was the muffled sound of an approaching car. The Italians had heard it too, but the officer pointing a gun at the POWs kept his forefinger on the trigger, tapping it lightly. The car sped towards the dump and stopped next to the POWs awaiting execution. A German officer stepped out and the Italians saluted him. Chanan understood that the newcomer outranked his captors.

Waving away the Italians, the German turned to the prisoners. 'Anybody speaks English here ... '

Chanan stepped forward and nodded.

'You're NCO . . . never work, only supervise, right?' the German officer asked.

'Yes, we are senior NCOs . . . not supposed to work, only supervise,' said Chanan. According to the Convention, NCOs who are POWs shall only be required to do supervisory work.[7] The German nodded, and Chanan, who was on tenterhooks, had to pinch himself to believe it. He was still breathing hard. The German officer had appeared as a saviour! It was a miracle.

The German ordered the POWs to be sent to a camp at Sirte, a city located east of Benghāzī, where the next afternoon, they received instructions to gather. Two Indian visitors waited for them. One wore a natty safari suit with a turban, while the other was draped in a white kurta and dhoti. They were from the Indian National Army (INA), which had aligned with the Germans in its efforts to free India from the British.

Amid the tumult of war, these gentlemen stood out in their impeccably crisp attire.

'Hello, we would like you to join our struggle to free India from British rule,' said the man in the safari suit.

'Has the German government agreed to air drop us on Indian soil to join the freedom struggle?' Chanan countered.

The visitors didn't appear trustworthy and pointed questions from the prisoners seemed to make them uncomfortable. It appeared the POWs had missed their chance to get away from the camp.

Later that day, Chanan and his fellow captives were taken to another camp at Tripoli near the harbour. Evidently, their fate had been sealed after the conversation.

A ship brought down by friendly forces

Meanwhile, fortunes changed quickly on the warfront.

The defeat and capture of Chanan's unit was the last of Rommel's victories. By mid-July 1942, Rommel had been contained at El Alamein. After a brief pause in the fighting, on 23 October 1942, the British 8th Army attacked from El Alamein. Rommel's forces were outnumbered and outgunned, and British forces started to gain the upper hand again. The tide had turned.

One morning, an excited prisoner burst in with news he had overheard. Japan – a German ally – had bombed Pearl Harbor, an American outpost, triggering talk of American marines heading to the ports of North Africa's Tunis and Algeria even as the Allied naval forces became active in the Mediterranean. Chanan was hopeful that the direction of the war might change, but he also feared its escalation.

The Germans, hard-pressed to ensure the custody of prisoners, kept moving their camps. Having lost ground in Africa, preparations had been under way to move the POWs to Italy by sea. At midnight on 9 October 1942, Chanan was among the four hundred British Indian Army POWs on board the Italian steamer *SS Loreto* crossing the Mediterranean Sea from Tripoli and headed to Naples, Italy.[8]

The concept of 'prisoner of war' encapsulated a classic absurdity within a conflict. Being taken prisoner stripped soldiers of their ability to engage in combat on battlefields and yet kept them in the war. They lived among the enemy, whom their allies strove to destroy. Death on a battlefield was honourable and preferred to being bombed by friendly forces. Chanan's ordeal was about to take a new direction.

On the night of 13 October 1942, at around 1630 hours, a deafening noise tore through the air startling the POWs who had assembled for evening prayers. Moments later, the ship shook and began to sway. Seawater broke through the sides and flooded the lower level. Their vessel had been hit.

A commotion ensued on the lower levels of the *Loreto* as the POWs rushed to save themselves. Chanan ran towards the only ladder that led to the deck. The other ladder collapsed under the weight of the many prisoners who tried to climb it simultaneously, setting off a stampede. Through the jostling, a few prisoners managed to escape. Many remained trapped in the basement.[9]

As Chanan reached the deck, a scene of chaos unfolded before his eyes. The captain's voice pierced through the tumult, barking urgent orders to abandon ship.

While sailing through the Tyrrhenian Sea, the *Loreto* had been torpedoed by the British U-class submarine *HMS Unruffled*.

Survival became Chanan's sole aim at this point. Observing that the Italian guards and crew were being issued life jackets, he and a few other Indian prisoners hurried forward to ask for theirs. The Italians turned them away. A scuffle broke out and the Indians managed to grab a few jackets. Hierarchy dissolved in the face of death.

Time was running out. Wearing life jackets, Chanan and fellow prisoner Mohammed 'Hafiz' Shafi leapt into the sea. Other POWs followed suit. The non-swimmers made a desperate attempt to grab Chanan and Hafiz, who may have drowned had they not fought them off. However, the sinking ship's chimneys heated the seawater to a dangerous temperature, resulting in horrific consequences for those who drifted towards the bubbling water – they were burnt alive.

Gasping for breath, Chanan and Hafiz grabbed a broken wooden plank, each gripping one side as the occasional unruly wave threatened to toss them off.[10] The boilers and chimneys of the ship had started to go under. They watched as the ship listed, held for a moment and toppled majestically into the fiery abyss of the sea. The soldiers trapped in the second basement had sunk with the ship.

The sea had turned rough by the time the steamer sank; it was dark and terrifying. The survivors experienced their life jackets coming loose, with metal buckles stabbing painfully at jaws and collarbones as they bounced on unruly waters.[11] After about three and a half hours, Chanan heard the roar of airplanes overhead. They flew past them and headed in the northwest direction. By now, the wind was picking up fast. Hafiz's head was drooping but his eyes were wide open. Without any sign of the shore, his energies waned. As the waves surged high and buried them in the foaming sea, his grip on the plank loosened.

Meanwhile, Chanan let his body float, conserving energy. As he bobbed in the water, another seven-foot wave pulled them under. When the wave receded, Chanan found himself winded, sucking all the air he could, gulping water and bouncing ungainly on the choppy waters – having lost complete control of himself.[12]

'Hafiz! Hafiz!' Chanan called out.

Hafiz had disappeared under the water. He was gone; now, the sea had calmed too. Chanan could hear the gushing sounds of water, but the biting cold had frozen every sinew and muscle in his body. He had lost strength in his arms, and he could shout no more. For the first time since their ship had gone down, Chanan felt alone in the vast ocean.

The sea had taken Hafiz and was coming for him next.

But no. He must survive, he reminded himself.

As Chanan struggled to stay afloat, he saw a tiny spark in the distance. It was a storm boat with a flashlight. Even as the seawater burned his eyes, he spotted a few rescue boats looking for survivors.

Chanan was ready to surrender to the force that was pulling him under. Was this the end?

The next thing he realized was that he could no longer feel the sea beneath him. His body was at rest. He was on a boat.

When his rescuers had reached him, Chanan was in a semi-conscious state, and they had struggled to pull him up.[13] He had placed a foot on the gunwale (the upper edge of the side of a boat) a few times, but his hands had failed to find a grip and he had tumbled back into the sea again. His rescuers made several attempts before they succeeded in retrieving him.

He had been found just off Capo Gallo in Palermo, Sicily. On the rescue boat, Chanan lay quietly staring up at the stars shining brightly in the clear, dark sky. He felt completely drained. The salt from the water edged his lips and his eyes stung as he struggled to open them. It was so peaceful that he could hear the susurrations of the waters whispering around him. The silence gave him a moment to think of his friends who had gone down with the ship. And what of Hafiz? They had made it out together, but Chanan would never see him again.

There were soldiers around him now; it wasn't surprising since the war was everywhere.

He picked up their German accent. 'Had they been late by a few minutes I would have also met the same fate as my other comrades,' Chanan thought. Chatting among themselves, the soldiers occasionally glanced his way. 'What fate! Our rescuers are German, and the ship was torpedoed by a British submarine.' Chanan smiled as he thought of the irony. One of the German soldiers smiled back. 'Whose war was it anyway?' he pondered as he slipped into unconsciousness.

Next, Chanan felt a clean sheet beneath him. There were no sounds of the sea any more. It was morning, yet he could barely open his eyes, and his arms and legs ached.

A woman sat in a chair beside his bed. She greeted him with a gentle smile and held out a bundle of clothes for him to wear. Startled, Chanan realized he was stark naked and quickly covered himself with his blanket. The woman was a nurse.

His turban was missing. 'I need to cover my head,' he told the woman. After some deliberation, they gathered bundles of dressings and fashioned them into a makeshift turban.

After coffee and breakfast, a guard appeared. 'Where are you taking me?' Chanan enquired.

'Sicily Island.'

After ten days in Sicily, a boat took Chanan across the sea to mainland Italy. A prisoner's camp at Udine awaited him.

The sinking of the *SS Loreto* stands as an overlooked tragedy, where a British submarine inadvertently took the lives of its own soldiers. Among the casualties were 130 British Indian Army POWs who were on board and had fought for the British Army in the war. Chanan had survived the battle at sea, but the war wasn't yet over for him.

The harrowing saga of escape and recapture

Prigione di Guerra 57 (Prison of War 57) in Udine was a prisoner's camp on the north-eastern coast of Italy. Chanan joined the camp that held American, British, Australian and Canadian prisoners. There were a hundred-odd Indian POWs too.

A British prisoner had assembled a radio receiver, which the other inmates helped conceal. In turn, the device allowed them all to monitor the progress of the war. One day, a Canadian NCO heard some good news; relations between German and Italian leadership were at breaking point as the latter had lost interest in the war.

The beat of the sentries, who kept guard inside camp, changed every hour and a half. As stories of soldiers deserting began to circulate, Chanan observed a lack of vigilance in the camp security as well.

The prisoners saw an opportunity and decided to escape.

A tunnel was planned from the farthest cell, which was 15 feet away from the perimeter barbed wire fence. Three dozen prisoners volunteered to work on the escape plan and gathered several essentials from the camp. Sharp-edged iron rods, a small 'crowbar', big spoons, tins of sorts and sharp-edged objects made up the stocks. The wooden floor of the barrack was 2.5 feet above the ground. Dislodging the wooden planks that formed the floors, the prisoners set about the task. Two planks were taken out at a time and the digging team slid under to work on a tunnel. Each team worked for forty minutes after which they were replaced by another. A prisoner designated as a 'holy man' – entrusted with lying on the bed and meditating as the others worked underneath – was a scout for the diggers. He warned them of threats and decoded messages from the tunnel-makers under his bed.

Crawling, digging and chipping away into the soft earth, the men completed the tunnel in seventeen days. When the order of escape was announced, some of the prisoners backed out, scared of capture as it meant certain death. On the first day, seventeen POWs went into the tunnel at 0220 hours and, after an hour of painstaking effort to open the tunnel exit, successfully escaped. Chanan was the last one out. The exit was meticulously covered by the foliage of bushes.

The escaped prisoners headed in different directions. While the white men found it easy to pass off as locals, Chanan stood out conspicuously among the Italians. In the meantime, a camp guard had noticed a drop in POW numbers and raised an alarm. Patrols began their search efforts and Chanan, who had been hiding in the streets, was forced to alter his plan and seek shelter indoors.

On day three, a kind-hearted Italian family offered him refuge, hiding him in a small storage area at the back of their house and even providing him with a meal. Late in the evening, there was a knock on the main door and Chanan could hear Italian soldiers speaking with the family members. 'It's a routine search,' they said. One of the soldiers wandered around the house. Just as they were about to leave, the wandering soldier kicked open the storage room door in the courtyard. It swung open and hit Chanan, who was hiding behind it. The soldier and POW faced each other; for Chanan, it felt like déjà vu as he found himself in a familiar situation once again. The soldier pointed his rifle at him, ready to pull the trigger. He raised his hands in surrender.

After searching his pockets, the soldier decided to spare him.

Chanan was captured and brought back to the camp, where he was sentenced to twenty-one days of rigorous imprisonment. He was thrown into a low-ceilinged, six-by-six-foot single cell with a canister in one corner for his ablutions. Life in solitary confinement was lonely, harsh and demeaning. Food was served once a day and he had to eat amid the terrible odour from the canister.

Upon his release from solitary confinement, Chanan was kept under strict surveillance. Prisoners were banned from speaking with him.

At this time news arrived of Mussolini being arrested by the Germans and their having taken over northern Italy. Chanan, who had been to the major theatres of the war – Africa, the oceans and Italy – could sense another move on the horizon. He was right; POWs from Italy were being transported to Germany even as it fought the war with its back to the wall. The Udine camp was hastily evacuated, and the POWs were moved to Stalag 12A in Limburg, western Germany, via a rundown goods train.

Stalag times: Prison camp in Germany

'If you have survived an ongoing war, you are either engaged in fighting the enemy or have been taken as a prisoner,' Chanan thought, pondering over the irony of the situation.

On a bitterly cold midnight in December 1942, Chanan joined Indians, Russians, Americans and British POWs at the large transit camp for newly captured prisoners. The Limburg stalag ('stalag' is camp in German) was notorious for its lack of basic amenities. There was no electricity, doctors or medicines. Six people crammed into a small bunker on hard cobblestones, arranging a few scattered straws for a makeshift bed. The living conditions were terrible, with diseases like diarrhoea spreading, and the air reeked of faeces, vomit and urine. As more prisoners arrived, the camp became a filthy cage.

The cage had its ground rules, though. A German soldier read them out on arrival. 'Do you see the tall barbed wire fence around the camp?' Chanan and the other POWs fixed their gaze on the spot the German soldier was pointing to. 'If you ever place your finger on it, you will be shot.'

For many new prisoners, Stalag 12A was their first opportunity to write a postcard home. It would take weeks for their missives to reach, but families whose loved ones were listed as missing in action celebrated when postcards arrived, bringing news that they were alive.

Chanan at a wrestling bout at Stalag 12A camp

In mid-1943, an intriguing development unfolded at Stalag 12A as ex-POWs who had joined the INA at Sirte arrived in the camp.

Chanan learned from these men that INA founder Netaji Subhas Chandra Bose had left Germany and flown to the Far East. Bose had proposed raising and dispatching two legions to North Africa to help German forces, but the plan was scrapped and the legions were disbanded. As a result, the veteran Indian soldiers with North Africa experience had been sent back to the POW camp[14] (see map on page 83).

Chanan was a leader at the prisoner camp. Conducting a parade.

The final chapter: The end of the war

At Stalag 12A, news of the war trickled in. The Allies were now making progress, pushing back the Germans, but being in a German prisoner camp, Chanan and the other POWs remained unaware of these setbacks.

POW Camps in Germany Including Stalag 12A

Stalag 12A

But Chanan had noticed that American bombers were dominating the airspace, unchallenged. They arrived in droves and pulverized the area frequently. The irony persisted – Chanan and his fellow inmates now kept a lookout for American bombers, ducking alongside German prison officials, and helped each other survive the bombardment by American planes. Indians, French, English and Americans were now in hiding alongside the Germans.

Sighting an airplane circling overhead, Chanan had once told his German captors to take shelter. 'Hey . . . hide there, you!!' he yelled. Another time, he pulled one of the German guards into the bunker, saving his life. A German guard similarly pushed a couple of English prisoners out of harm's way when a bomb fell in the open courtyard. Once, the French section was bombed, causing deaths; English prisoners fled their own countrymen's bombs on the playground another time. The prison camp had become a home to captors and prisoners alike – and they had to protect each other to survive.

The Americans were winning the war. From his trench, Chanan could see the bombers in formation, emerging out of the skies – tiny at first, followed by a menacing roar that ended in a ferocious bout of carpet bombing. There were two hundred raids every day, every few minutes. The attacks were relentless. Everything on the ground was blown to bits. Those in the trenches were forcibly ejected with pinpoint bombing. No place was safe.

Had Chanan come this far to die at the hands of his own military? It was a thought that troubled most POWs at the camp.

By June 1944, news arrived that Allied troops had landed in Normandy in France. A bold and upbeat Chanan asked the barrack commanders to read out the news to the other POWs during roll call. Shortly after, Chanan met a British brigade commander who had been captured. He couldn't help but notice a twinkle in the British commander's eyes. 'Is there something you want to tell me, sir?' asked Chanan.

The officer told Chanan that Allied forces had finally defeated the Germans. Chanan stood there, initially numbed by shock and disbelief, but soon he was overcome by elation. Tears rolled down his cheeks as he ran to the Indian camp.

A pass for Chanan to travel inside Stalag 12A

Source: Gurbinder Singh Dhillon Personal Archives

'We are going to be free! Free! Free! Freedom . . . ' his words rang across the barracks. This was the biggest news of all, surpassing the defeat of the Germans and Allied victory.

He hurried up to his friend – one Sergeant Charles, an American prisoner who had taught at an American university before the war. Charles rushed back to his men and broke the news.

The POWs faced a lengthy wait ahead, enduring a particularly harsh December winter.

Every morning, Chanan would wake up and run to the British brigade commander. 'Any news, sir?' A shake of the head or shrug was a typical response.

By April 1945, the camp had been surrounded by Allied forces. One morning, Chanan saw the Germans abandoning their posts, shedding army attire and vanishing into the civilian population of the town. Chanan and his mates were now awaiting their saviours inside the prison camp. The fortunes of war had come full circle.

Morning broke with the rumbling of tanks and armoured carriers. Friendly soldiers with flags atop their tanks arrived from everywhere, waving at the prisoners who were now the sole occupants of the camp.

The evacuation started soon after. Many prisoners were weak, diseased, undernourished and unstable. Chanan had lost considerable weight and strength, but he kept himself busy, cleaning, bathing regularly and taking walks to improve his fitness.

His diary entry after his evacuation from the Nazi prison camp sheds light on that precarious time. 'On 2 April 1945, I am flown to Paris. There I stay for two weeks with American soldiers for rest and convalescence. Here I see the service

Chanan played sports such as table tennis to keep fit.

zeal among the nursing sisters and the medical staff. Though undernourished, I am in a reasonably fit state of mind and body, and I am able to do my chores of bathing, cleaning and other sundry jobs of day-to-day life. But whenever I do so, the sister on duty forbids me and does everything for me. Once I stood at the window to watch a football match going on outside next to the hospital premises. She promptly came and put me in the bed, so that I did not strain myself,' he writes.

Chanan was flown to London where he awaited his repatriation to India. There, prisoners were separated into groups, as the British were targeting INA members. These INA enlistees had initially chosen to join the organization but were later transferred to a POW camp in Germany after the alliance between Bose and Hitler disintegrated.

The INA soldiers were identified by the British as traitors to the Raj. Surprisingly, both the post-independence Indian government and the British government, once adversaries in the fight for freedom, shared a similar perspective on the INA, leading to their subsequent isolation.

India wins freedom. Chanan waits.

After India attained freedom on 15 August 1947, soldiers from WWII were absorbed into the new Indian Army. Chanan Singh Dhillon's case, however, was unique. He had waited the entire war to become a commissioned officer. A new, independent Indian army – saddled with the burden of history and bureaucracy – would ensure the wait was longer.

After returning from internment at the POW camp in Germany, Chanan rejoined the army. He was posted in Rawalpindi till the end of 1947, when a bloody partition took place dividing British India into two countries. Indian army units were withdrawn from the western regions that now fell under the jurisdiction of Pakistan. Chanan was sent to Ambala, and later Kashmir, as trouble started brewing between the two new nations: India and Pakistan.

In this new country, beset by partition, violence and the war of 1948 between India and Pakistan, the letter of commission that would have made Chanan an Indian Army officer was forgotten.

In 1957, a decade after Independence, Chanan received a call from his adjutant in Rajouri (in present-day Jammu & Kashmir). The adjutant, an officer assisting with administrative duties and coordination in an army unit, had good news. Chanan's application papers for officer training had been approved. He was summoned by General Shiv Dev Verma, General Commanding Officer of 15 Corps in Kashmir, who grilled him for an hour. Impressed, the General told him to prepare for his selection board exams.

Upon returning from the war, Chanan had married. On 27 November 1960, while on his posting, he received news from the hospital. His wife had given birth to a son. Chanan smiled; he saw it as a happy harbinger of things to come. Later that day, he finally received the news he had long awaited. He had been selected for training to be a commissioned officer.

It was almost as though that ship from Egypt, after all these years, had finally come home.

On 31 October 1975, Chanan Singh Dhillon retired from the army as a lieutenant colonel. Until his peaceful passing in 2011, he had come to terms with the challenges life presented him. 'Thirty-seven years and 26 days, full of ups and downs. I dedicated all my energy and youthful days to this service. War, hunger, hardships in the desert and at sea, captivity and freedom. My commission as officer helped transform my children's lives through the efforts of my wife who stood by me through thick and thin,' read his diary entry.

Postscript

On that trip to Limburg in 2018, Chanan's son Gurbinder went to the place Magyar had shown him on the map. However, it was different from what he had imagined! Instead of the signs of a Stalag 12A Nazi prison camp, he found a police training school and refugee camp on the site.

His next stop was the town hall office. Spotting a clerk

who spoke English, Gurbinder asked him what happened to the place he had come looking for.

'Oh, we have preserved the remnants from those times. We wish to make a memorial some day . . .' said the bored clerk.

Gurbinder sighed, pursed his lips and looked around the busy office. Everyone seemed engrossed in the present; perhaps, he was the only one looking back at the past. Maybe he was standing where his father had once stood, proclaiming 'freedom' many years ago.

A disappointed Gurbinder returned to India, with a fistful of earth from Limburg. He was consoled by the thought that his father, Chanan Singh Dhillon, went on to live a full life. Many lives, in fact. And it all started the day he received that life-changing letter, offering him the opportunity of his dreams.

Part II

Defending the Borders

—•————————•—

In the wake of India's independence in 1947, a series of conflicts unfolded in the neighbouring regions and soldiers engaged in wars to protect the homeland or redefine borders.

Stories of loss, redemption, resilience and triumph inked human relationships as soldiers faced adversaries closer home. Bittersweet stories arrived from the borders about notions of nationhood in a troubled neighbourhood.

—•————————•—

4

The Boy Who Would Become Stak

The Legend of Chhewang Rinchen

On 21 October 1947, five thousand Pakistani raiders stormed Kashmir aiming to capture the state from India by force.[1] These raiders, consisting of both tribal fighters and regular Pakistan Army personnel, advanced towards Srinagar hoping to seize it. Their incursion marked the beginning of war in Kashmir.

Under pressure, Maharaja Hari Singh of Jammu & Kashmir signed the instrument of succession (transfer of power) for Kashmir in favour of India. Colonel Lionel Pratap 'Bogey' Sen was hurriedly made a brigadier and dispatched from New Delhi to command 161 Brigade with a task to drive out the raiders and secure Kashmir.[2] On 27 October, troops from the 1st Sikh Regiment of the Indian Army were airlifted from New Delhi to Srinagar to counter the invasion.[3] Launching a fierce counterattack against the raiders, they successfully pushed them back.

Ladakh, however, remained under threat.

The mountains of Ladakh were at a considerable distance from the Kashmir Valley and challenging to reach. So, the Pakistan-aided raiders approached these mountains from the west, leveraging the advantage of the mountainous terrain and the proximity of the Pakistan–Ladakh border. Their campaign had been going to plan.

Meanwhile, the Gilgit–Baltistan region, part of the erstwhile princely state of Jammu & Kashmir, saw a rebellion by the Gilgit Scouts.[4] This paramilitary force, composed of local men and formed by the British, opposed the decision of the Kashmir royalty to align with India. They sided with Pakistan and, along with local militias and tribesmen, actively engaged in numerous skirmishes and battles throughout the conflict. It was largely due to their efforts that Gilgit fell into Pakistani hands in November 1947.[5]

By February 1948, Ladakh was under siege by enemy forces. The northern front of Kashmir was being opened up. A small garrison at Skardu had been surrounded.[6] The towns of Gurez, Drass and Kargil were under threat.

To attack Leh, the capital of Ladakh, the raiders had several choices for routes from north, west and south. The weakly held garrison at Leh wasn't expected to offer much resistance. The snow had temporarily aided the defenders, but their situation remained precarious.

While the Indian army was busy fighting the raiders in the valley, Brigadier Sen dispatched a small detachment of 20 Lahaulis and Ladakhis led by Major Prithi Chand of 2nd Dogra Regiment to Leh. Prithi Chand was tasked to raise a force capable of defending the city.[7]

Prithi Chand and his unit crossed the treacherous Zoji La pass – connecting Srinagar and Leh – in winter to reach Leh on 9 March 1948. On arrival, he found himself in a tough spot. Only thirty-three men from the State Forces were available to defend the city against thousands of attackers from Pakistan.[8] Prithi Chand needed volunteers until reinforcements arrived from Kashmir.

Meanwhile, Leh hoped its meagre defences would delay the advance of the Pakistani force till the reinforcements of the Indian Army arrived.

After Prithi Chand and his men's arrival in Leh, the Indian tricolour was raised, and the British residency flag was lowered. On the historic day of 13 March 1948, outside Karzoo Palace, the erstwhile British Residency in Leh, an enthusiastic crowd belted out the Indian national anthem to the sound of Buddhist cymbals and drums.[9] The reassuring sight of Prithi Chand and his men instilled a sense of hope, providing a much-needed respite after the months of darkness cast by the Pakistani raids on Kashmir.

'Ki ki so so lha gyal lo, Hindustan Zindabad (Victory to the gods, long live India),' people chanted.

Prithi Chand stood on a raised platform and surveyed the assembled crowd.

'Koi hai? Koi volunteer hai (Anyone willing to volunteer?)?'

His question was met with an anxious silence.

'We have come to protect your chorten and the gompas,' he said, appealing to their instincts to safeguard the Buddhist religious structures.

Still, no one spoke. After assuring them reinforcements would come, he pressed on. 'Anyone, anyone?'

A moment later, a hand in the back shot up.

Chhewang Rinchen was the first soldier to volunteer to fight for his homeland that morning. He was 17 years old.

Part I: Leh, 1948 – Enemy of the gates

In recognition of his instrumental role in repelling a Turkish invader, the Gyalpo (king) of the land had bestowed upon one of Chhewang Rinchen's esteemed ancestors the title of 'Stak' (Tibetan for tiger).

Imitating his illustrious forefather, little Chhewang had wielded a stick like a sword. He swung it around in the make-believe battles he fought while growing up in Sumur village in the Nubra Valley, north of Leh.

'Mother, when I grow up and fight the invaders, will you then call me Stak?' he had asked, and his mother had blessed him.[10] Years later, raising his hand in response to Prithi Chand's request made in 1948, Chhewang took his first step towards achieving this goal.

May

By May 1948, the snow had melted, clearing the way for the raiders who were intent on swiftly annexing Leh.

By the end of the month, they had managed to capture the towns of Kargil and Drass and the crucial Zoji La pass, cutting off the line of communication between Srinagar and Leh. There were no Indian troops between Skardu and Biagdangdo village (now Bogdang). Reports arrived that a column of raiders had reached Biagdangdo, located in the Nubra Valley. On 22 May, the State Forces at Khalatse (also known as Khalsi) were attacked by the Pakistani raiders and defeated.[11] As the troops retreated after the defeat, the State Forces platoons demolished the bridges, halting the enemy in their tracks and delaying the advance.[12] This was important since Leh was a little under 100 kilometres away: a day's march from there.

The threat to Leh was at its gravest.

The Indian Army was still occupied in securing the Kashmir Valley, after the Pakistan Army raiders were beaten back. Chand's unit was ready, but their numbers were sparse.

A slim chance of saving Leh seemed to be through the airlift of troops, similar to what had been done in Srinagar in October 1947. Sonam Norboo, an engineer and Ladakhi native, had arrived in Leh with Prithi Chand to make an airstrip in the region. The available makeshift runway was small, and it was deemed impossible for a Dakota aircraft, which was of moderate size, to land on such a restricted airstrip.

That is, until a daredevil pilot proved them wrong. One afternoon in Srinagar, on the banks of the Jhelum river, over strawberry and ice cream, Air Commodore Mehar Singh and Major General K.S. Thimayya had furiously debated a solution to the crisis. Thimayya, the General Officer Commanding (GOC) of the 19th Infantry Division, responsible for the defence of Kashmir, urged Singh to provide air support to relieve the siege of Leh. Singh had been sceptical. 'Can an aged and battered Dakota with a war-worn engine be trusted for a flight over treacherous ranges of mountains along a hitherto uncharted route at a height of 6,000–7,000 metres?'[13]

Thimayya had a solution ready. He offered to fly along with him. 'Let's go,' he ordered. On 24 May 1948, Singh accomplished a formidable undertaking of landing a Dakota on one of the highest airstrips in the world.

Thimayya promised Prithi Chand they would send reinforcements, but poor weather conditions caused some

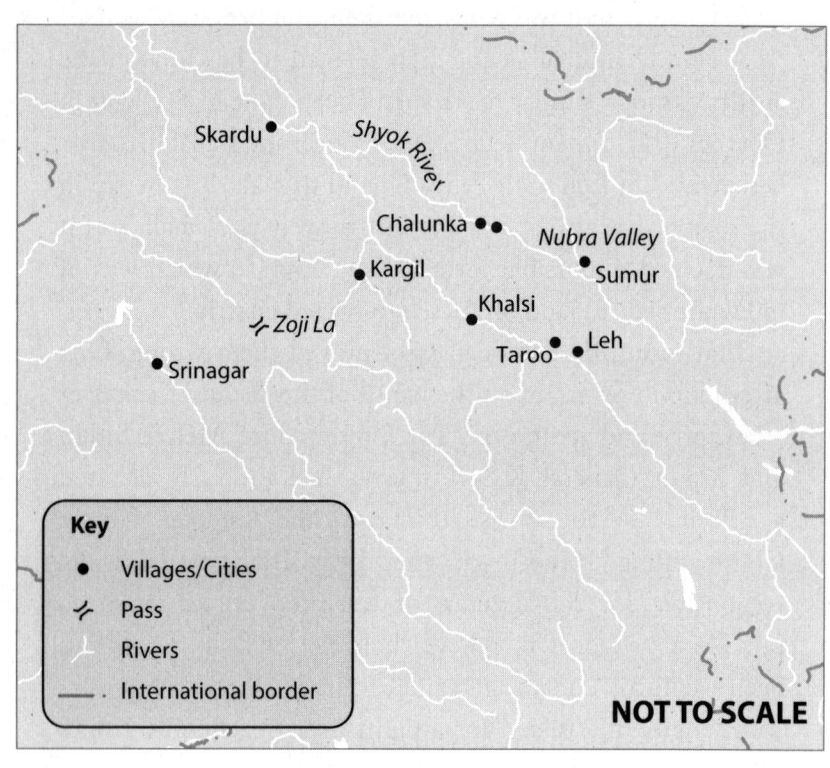

Places in Ladakh in the Story

delays. However, Singh soon led a flight contingent consisting of a company of 2/4 Gorkha Rifles, along with essential supplies and ammunition, to fortify Leh[14] (see map on page 98).

June

Raiders continued to pose a threat to Leh, and one of the most unguarded routes to the city was via the Shyok river and Chhewang's village (see map on page 100).

After undergoing ten days of training with Prithi Chand's second-in-command, Subedar Bhim Chand, they were tasked with raising a local force in the Nubra Valley. Carrying a Sten gun, Chhewang looked every inch a guerrilla fighter. Prithi Chand thought he had found not only a volunteer but also a leader.

The villagers Chhewang appealed to, however, seemed unconvinced that a teenager could lead a defence of the city. Frustrated, Chhewang made an emotional plea. 'If you don't help, the land will be captured, idols of the Buddha broken. Come on . . . we risk the security of our women, our homes, everything. Are you prepared for the looting and burning? Are you ready? Tell me . . .' he implored.

Stanzin Tsering, a headman from the Nubra region, was moved enough to assist him. Soon, twenty-eight volunteers joined Chhewang and the Nubra Guards was born.[15]

The Nunnus, a nickname given to the Nubra Guards, were put through a short course on weapon craft and combat training by Bhim Chand. Once the instruction ended, they initiated a successful campaign to push back the intruders.

On patrol, the Nubra Guards reached Chumik La, a mountain pass that is a gateway to the valley and came upon

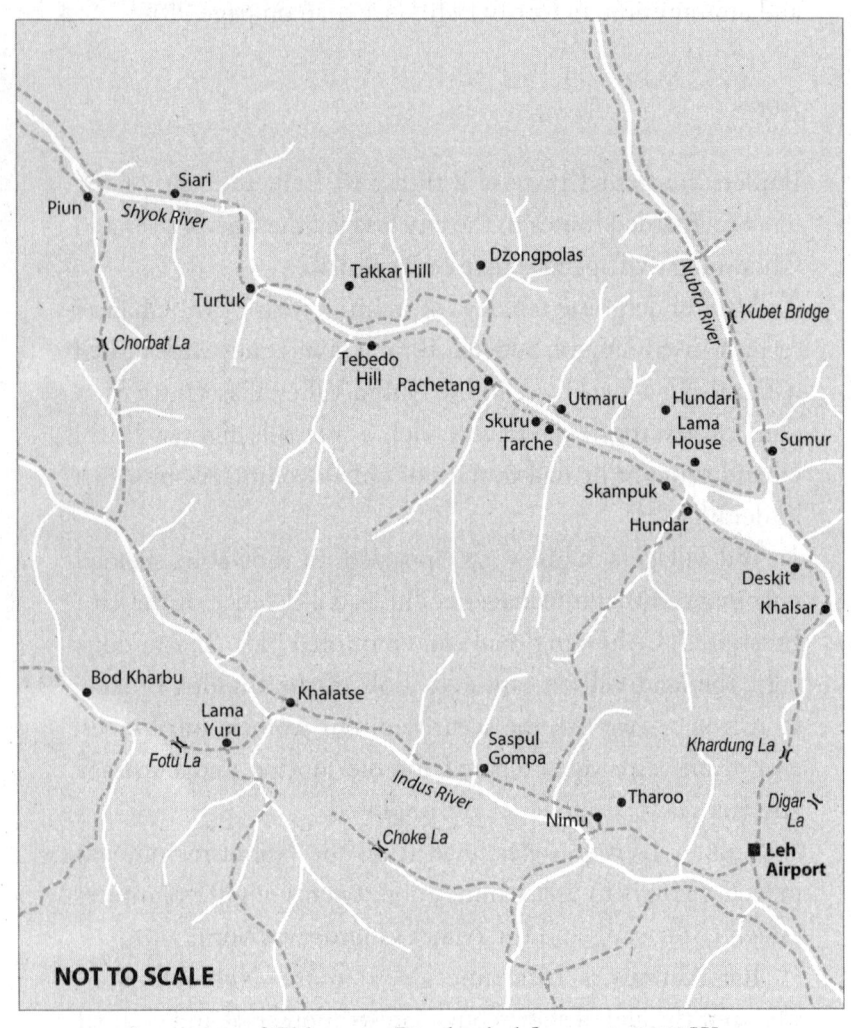

Locations of Chhewang Rinchen's Advance in 1948 War

weary Pakistani forces resting at their post. Caught off guard, the raiders were swiftly defeated, resulting in the capture of the Pakistani post and the confiscation of weapons, with two raiders killed.

Chhewang's men had had a successful start.

However, the situation at Leh was constantly evolving. Soon after, anticipating an enemy attack at Leh, the garrison issued orders for all the forces to retreat to Leh to protect the capital. The order stated the Nubra Guards was to be disbanded, their arms and ammunition were to be withdrawn.[16]

The news created panic among the people. The absence of the Nubra Guards left the Nubra Valley exposed to Pakistani raiders. The next morning, panicky villagers lined the roads heading for Leh. Little did they know that the capital was also under threat.

Chhewang watched as the groups left; it was a futile exodus. He met with the prominent members of the region, including Stanzin Tsering, who had backed the recruitment campaign a few months ago. They urged him to raise forces to safeguard the region and keep the raiders at bay.

The road ahead was tough and uncertain. At a nearby temple, Chhewang prayed for divine intervention before Lord Buddha. On the Buddha's visage, he sensed a serenity amid the maelstrom.

A few days later, Chhewang met Prithi Chand in Leh and argued his case with passion and confidence. He insisted on being given weapons to defend against the enemy.

Prithi Chand had scarce resources and was facing a rough time in Leh. He pondered the potential drawbacks. What if the weapons were confiscated? What if the Nunnus were captured?

On the other hand, Chhewang exuded new-found confidence and a strong sense of purpose after his first successful raid. 'I need them, please,' he said firmly and, eventually, Prithi Chand acquiesced.

Chhewang prepared to head back to the Nubra Valley with 28 rifles, one Sten gun, one box of hand grenades and three boxes of ammunition. He hired a pony from Leh to carry the weapons. As Prithi Chand watched him leave the office, his confidence in Chhewang grew. That day, he felt the young boy was destined for great things (see map on page 100).

July

Upon Chhewang's return to the Nubra Valley, locals from the village of Skampuk told him that, facing no resistance, the Pakistani raiders forced residents of the nearby Partapur village to arrange for 200 riding ponies, 100 loading ponies and 300 porters. The villagers instantly knew that fresh Pakistani forces were being summoned for their march to Leh. Additionally, news arrived that the enemy intended to cross the Shyok river by boat (see map on page 106).

Noting this approach as the perfect 'guerrilla killing ground' due to the narrow landing area that would naturally result in the bunching of Pakistani soldiers, Chhewang and his men decided to ambush the enemy.

At the break of dawn, a few zhaks (rafts) lined up along the banks of the river. Newly arrived Pakistani soldiers quietly boarded the zhaks. On the other side of the river, Chhewang and his men took up positions with their rifles. Chhewang had instructed his team to let them land on their side of the bank and then select the targets.

As the zhaks reached mid-river, an excitable young Nubra Guard pulled the trigger. The outcome was immediate: a couple of soldiers fell backwards into the river. Others jumped in. With the element of surprise lost, Chhewang's men swiftly attacked the soldiers crossing the river, unleashing a barrage of gunfire. In response, the soldiers in the water retreated, while their colleagues on the other side opened fire at the Nubra Guards.

After a sustained exchange of fire, the enemy eventually retreated towards Hundri village. The Nubra Guards had foiled their plans.

August

Prithi Chand's predicament in Leh, on the other hand, had been deteriorating rapidly.

After the 2/4 Gorkha Rifles were airlifted into Leh in May, a detachment of 2/8 Gorkha Rifles led by Major Hari Chand had arrived from Manali on foot.[17] However, the raiders successfully overwhelmed the Indian defences and advanced to Taroo village, only 16 kilometres away from Leh. By the end of July, Prithi Chand's ammunition supply had dwindled further.

It was at this time that Skardu fell into the hands of the enemy.

Under siege from Pakistani raiders and isolated from their own troops since February, Lieutenant Colonel Sher Jung Thapa and the State Forces had valiantly defended the Skardu garrison for months. Their courageous efforts had proved crucial in effectively delaying the enemy's advance towards Leh.

When the defending forces, having fought valiantly, were

eventually defeated on 14 August 1948,[18] the raiders unleashed a horrific wave of violence. Men, women and children suffered tragic fates, while the small contingent of Indian soldiers were brutally executed.

The fall of Skardu and its aftermath sent shockwaves through New Delhi. The question arose: Would Leh face a similar fate?[19]

Things now began to move rapidly.

Another company of 2/8 Gorkha Rifles, led by Lieutenant Colonel H.S. Parab, was airlifted to Leh while the remaining battalion embarked on foot via Rohtang Pass. Before being airlifted, Parab was appointed as the Military Governor of Ladakh.[20] Under explicit orders from Thimayya, who regarded Skardu, already fallen into enemy hands, as the final frontier in the battle to safeguard Ladakh, Parab had been instructed, 'You will defend Leh at all costs.'[21]

Soon, Leh was thronging with troops, forming a formidable military garrison worthy of the name. Prithi Chand, who had been fighting a solitary battle until this moment, smiled at the sight.

September

The massacre at Skardu catalysed the region to action.

With the arrival of reinforcements in Leh, more young men joined the Nubra Guards. Armed with swords, spears and muzzle-loading guns, they carried provisions such as sattu (barley), flour, meat, apricots and chhaang, a Tibetan alcoholic beverage. A resolute fighting force, these men were committed to defending their land at any cost.

In the Nubra Valley, intelligence sources in the village

brought news that Pakistani troops aimed to capture the young commander of the Nubra Guards, who was obstructing their advance. Chhewang sent a messenger to Prithi Chand, asking for reinforcements. Prithi Chand wrote back to his teenage protégé, praising him and his unit for their bravery and promised to send a battalion of 7 J&K Militia (later, it was converted to a full-fledged army regiment and called Jammu & Kashmir Light Infantry [JAK LI]). Chhewang also received formal enlistment as a Jemadar in the 7 J&K militia.[22] Building upon this role, he assembled an army of 100 men while an additional 100 unarmed included individuals who served as valuable support, working as information agents and aiding in weapon transportation.

Amidst the bustling preparations, Chhewang's mind recalled a cherished childhood tale of the renowned Dogra military general General Zorawar Singh. The story recounted how the unwavering determination of local residents had triumphed over Zorawar's formidable forces, resulting in the defeat of the invading general. To Chhewang, it appeared that the entire Nubra region stood resolute and prepared to confront the attackers, echoing the spirit of the legendary tale.

Additionally, Chhewang was tasked with raising a guerrilla company of Nubra Guards. He shortlisted fifty men who were trained to climb mountains from difficult angles, could survive in the snow for days and engage in combat with limited resources.

These men were assembled with the purpose of executing a grand plan.

Lama House, situated at a towering elevation of 4,500 metres on the opposite side of the Shyok river near Hundri, remained under the control of the Pakistani raiders. This

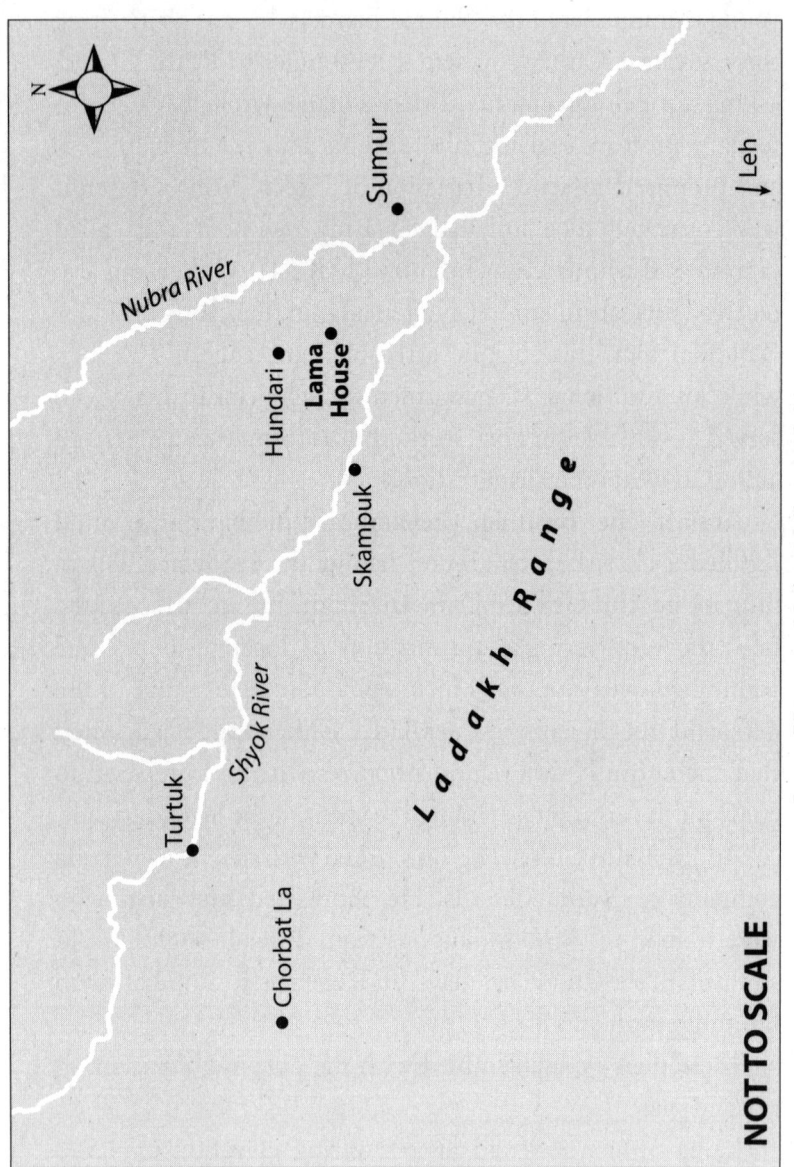

Lama House in Nubra Valley, Ladakh

strategic feature held the potential to cripple the enemy advance and derail their plans in the area, if seized (see map on page 106).

Chhewang's objective was to capture Lama House, and he set off on the mission with his guerrilla team. Enduring harsh weather and perilous routes, they finally arrived at the path along the Shyok river, which they had to cross. They decided to use the Kubet bridge.

The enemy had been on high alert in this location ever since their zhuks had been ambushed by the Nubra Guards in July. Both the bridge and Lama House were heavily patrolled.

The Nubra Guards stealthily crossed the bridge at night, their tracks masked by the murmuring of the river, and established a base near the target mountain.

Chhewang's methods were distinctive. He decided that they would climb to a higher adjacent peak and descend by rolling to launch an attack on the enemy post.

The enemy had been anticipating an attack from the lower southern axis, considering the northern route was tougher, steeper and higher. Chhewang, being aware of that, deliberately chose the more difficult route, to maintain surprise. That's why he had assembled a group of tough guerrilla fighters to accompany him.

At Nullah Chakri Chubab, below an adjacent mountain, he established a base. After two days of climbing the mountain, the group reached a height of 5,000 metres. They could see the Pakistani post below, armed and prepared.

That night, the men crawled down the mountain slope purposefully but cautiously. During the reconnaissance patrol, they discovered a Pakistani platoon, with approximately twenty-five soldiers, engaged in constructing a wall around the post. Chhewang observed that the approach route posed considerable challenges. Layers of ice made the ground slippery and treacherous.

The team was divided into three sections. The attack was to be launched at night, to take advantage of the cover of darkness.

On the night of the planned attack, the slippery terrain hindered their progress, slowing their advance. Using ropes as their guide, and every ounce of strength, they painstakingly made their way through the darkness, sometimes even slithering across gorges. At 0300 hours, they arrived close to the enemy post.

Hearing sounds in the dark, the sentry, whose duty was to keep watch, became suspicious. Nervously, he took aim and fired, uncertain of his target. The gunshot sound alerted others in the post.

The Nubra Guards may have lost the element of surprise, but by then they had come close enough to launch a frontal assault. Before the enemy troops could get into position, the attacking forces charged, and Chhewang swiftly hurled a grenade into the main bunker. The team immediately targeted the post's commander, Subedar Mota Hussain of Gilgit Scouts.

Following a short but intense battle, in which ten enemy soldiers were killed, the Nubra Guards captured Lama House. Chhewang and his men showed how a small, smart and unorthodox force could overpower a bigger and more powerful enemy.[23]

December

The Indian army had retaken Zoji La and Kargil by mid-November.[24] The swift advance of the Indian troops unsettled the Pakistani-backed forces, prompting their retreat from the Nubra Valley. In hot pursuit, the Indian army reached Hundri and Tarche.

When the Indian troops entered Baigdangdo, Chhewang observed that the village was predominantly populated by men. How odd, he thought. The sector commander directed Chhewang to locate the whereabouts of the women and children. He found them hiding in a nearby nullah, their faces covered with soot. Upon catching sight of Chhewang and the soldiers, the terrified women hastily fled in fear. Chhewang called out to them in the local language, and said, 'Come back . . . you're safe here.' He assured them that they would be protected and treated well. 'You'll be safe from the Pakistani raiders . . . trust me,' he said. Upon hearing a soldier speak their language, a few of the women paused and turned towards Chhewang. With a reassuring smile, he repeated in the local dialect, 'We will protect you . . . your honour . . . you're safe.'

Tukkar Hill stood as the last enemy stronghold in Leh tehsil (subdistrict), yet to be conquered. On 15 December, Chhewang and his guerrilla force were tasked with capturing Tukkar Hill, while the main force engaged in attacking the heavily fortified Black Rock feature.

The Nubra Guards had fought continuously, with little respite, for months now. Despite enduring frostbite during the previous operation, half of Chhewang's platoon persevered alongside him and began the march, crossing a snow-clad mountain towering over 21,000 feet.[25]

It was a decisive victory. Pakistani forces had already been defeated and pushed back in other areas, and the Nunnus faced a demoralized, leaderless enemy at Tukkar Hill. Many opted to flee in the face of the advancing Indian forces. That night, Chhewang and his men celebrated with a barakhana (grand feast with troops) using the rations left behind by the Pakistani troops.

A New Year's surprise awaited them. On 1 January 1949, while awaiting orders to march towards Baltistan, Richen received news of a ceasefire between India and Pakistan. More astonishingly, the Pakistani sector commander invited his Indian counterpart to lunch at their headquarters at Chalunka, a small village in the Shyok river valley.

Accepting the invitation, the Indian contingent arrived one afternoon for a shared meal. Amid the few hours of camaraderie, Chhewang's gaze swept over the surrounding villages, contemplating what could have been. This land, including Chalunka, could have been ours, he thought ruefully. Chhewang regretted that the time spent in the barakhana had allowed the Pakistani troops to slip away.

By the time the war ended, young Chhewang had come a long way from an enthusiastic volunteer responding to Prithi Chand's call at the Leh square. For his valiant act of confronting and defeating a significantly larger enemy force and reclaiming the lost territory, seventeen-year-old Jemadar Chhewang Rinchen was awarded the Maha Vir Chakra, becoming the youngest recipient to get this prestigious gallantry decoration.

However, for Chhewang, there remained unfinished business in Chalunka. Little did he know that history would present him with another opportunity years later.

The Boy Who Would Become Stak 111

Part II: Leh, 1971 – An unfinished business

Twenty-three years after breaking bread with the enemy during the ceasefire at Chalunka, Chhewang finally seized the second chance he had longed for.

In 1971, the long-simmering tensions between India and Pakistan finally erupted into a full-blown war. While the focus of attention was on the major conflicts in the eastern and western sectors, a lesser-known but tough campaign unfolded in the mountainous region of Ladakh. Unsurprisingly, a familiar face from the past, a former teenage hero, found himself once again amid the intense action.

After receiving a regular commission in the Indian Army, Chhewang had rejoined his former regiment and was once again training and organizing the Nubra Guards. Now a

Colonel Chhewang Rinchen MVC Bar, SM

major, he had achieved legendary status, renowned for his victorious exploits in every battle he had fought. His obstinacy and overconfidence often led to clashes with superiors over military discipline and protocols, yet his unmatched battlefield prowess earned begrudging recognition from his seniors.

As a seasoned major in his 40s, Chhewang now faced a formidable adversary – a younger, more agile enemy armed with advanced weaponry during the Indo-Pak War of 1971.

As part of their strategic plan, the Indian army aimed to capture Point 18402, which stood as the highest post ever attempted in a terrestrial battle.[26] This crucial objective was currently held by Pakistani forces.

The task of capturing it was assigned to a 400-men column of Dhal Force under Major S.S. Ahluwalia. Ahluwalia was tasked with sending a platoon to the rear of Point 18402 while the rest of the troops held the front route.

The Indians made slow progress in the rarefied air of the high Himalayas. Altitude sickness, breathing issues and headaches troubled even the toughest of men in these mountains, and a few non-Ladakhi soldiers dropped out as they climbed.

Chhewang led the Dhal Force group along the front route as part of this operation. Over the years, his unorthodox approach to combat in high altitudes often put him at odds with his seniors. Believing that weather and terrain were crucial factors in determining what one should wear or carry, Chhewang rejected the army steel helmet for balaclavas and ammunition boots for Ladakhi paboos,[27] known for their lightness and warmth. (Paboos, shoes of woven yak hair and pashmina, are best suited for the rugged terrain and harsh weather conditions of Ladakh.)

While climbing up towards Point 18402, Chhewang noticed Major Thapa of the Gorkha Rifles pausing to look up at the objective, which was far away. Chhewang offered him his bottle.

Thapa took a swig and asked, 'What's this? Arak?' (Arak is an alcoholic beverage.)

Smiling, Chhewang shook his head.

The officer took a second swig and turned to Chhewang, grinning widely in the cold. 'Ah, it's rum mixed with water!'

It was a drink that kept the spirits invigorated. With a guffaw, Chhewang exclaimed, 'That is our drink in these heights!' as he continued to march. His confidence in victory that day was unshakeable.

The group was travelling light, to conserve energy for potential combat on the mountain's summit. They carried limited ammunition and supplies, relying on securing 'complementary' blankets from the Pakistanis to ward off the cold.

As the first glimmer of sunlight appeared on the horizon, Chhewang was startled by the distant sound of grenade explosions echoing from a few hundred metres above them. Ahluwalia's team, approaching from the rear, had initiated an attack on the post at the summit. The Pakistanis returned fire and this exchange went on for about fifteen minutes.[28] Chhewang located a Pakistani forward post which was bringing down heavy fire. To distract the enemy, Chhewang and the Nubra Guards began shouting, 'Hands up and surrender, else you will be killed.' The diversion worked. By 0700 hours, the highest post in the world had been captured by Ahluwalia and his men.

On the morning of 8 December, amid the resounding war

cry, 'Ki ki so so lha gyal lo', Chhewang stood atop Point 18402, gazing upon the vast valley. The sight stirred deep emotions within him, evoking memories of his solemn oath to defend the land. In that moment, he truly felt like a Stak, on top of the world.

On 9 December, the Indian troops advanced towards the Chalunka defence complex, the same place where Chhewang had shared a friendly feast two decades ago. Now, Chhewang led an assault along this risky route with Thapa and two platoons of the Nubra Guards. One platoon from each company was kept in reserve.

There was only one approach, through a nullah, and they cautiously crawled through it, taking care to stay hidden. Amid the Indian mortar shelling, Chhewang pressed forward, anticipating a counterattack from the Pakistanis. Surprisingly, no retaliation came. As they approached the post, the enemy suddenly unleashed a barrage of mortars, machine gun and rifle fire, leaving Chhewang's squad pinned down.

Thapa's team, which had moved separately, managed to silence an enemy bunker and kill a few Pakistani soldiers. Naik Fateh Mohammad of the Dhal Force crawled towards a Pakistani medium machine gun (MMG) trench. 'Kafiron ko toh main khatam kar doonga, mujhe hathgole chahiye (I will annihilate the infidels; I need hand grenades),' Mohammad shouted. The remaining Pakistani troops were swiftly subdued using bayonets, without a single shot being fired by the Indian troops.[29] The men had demonstrated composure and skill in close-quarter combat.

A few Pakistanis staggered out of their bunkers, surrendering with raised hands. They were pounced upon by Indian soldiers. Chhewang, embodying the warrior's code of honour with humble grace, reprimanded his fellow soldiers, saying, 'Surrendered soldiers are our guests.'

After Chalunka was won, plans were made to capture Turtok and Thang, villages on the banks of the Shyok river. On 12 December, Chhewang sent out patrols which brought news about Pakistani defences and preparations. The advancing Indian troops moved under the cover of intense mortar shelling, keeping the Pakistani defences pinned down.

Around 2200 hours, the shelling ceased, enveloping the area in an eerie silence. Chhewang's troops surrounded the wing headquarters and the village, cautiously entering what appeared to be an abandoned ghost town.

A pair of cows watched as the uniformed men stealthily infiltrated the village. While testing the doors to the houses, they discovered they were locked; it appeared that the inhabitants had secured themselves inside.

Journalist Shubhangi Swaurup wrote in an article how Chhewang knocked on the door of a prominent house and speaking in the Balto language, said, 'I am Ali, the porter, a resident of Biagdangdo. I come with the Indian Army.'

The door opened slightly, revealing two men within. Chhewang aimed his pistol at them and three Indian soldiers pointed their bayonets. 'Please don't be afraid of us; only tell us the truth. Is there any Pakistani soldier hiding in your house?'

Ghulam Hussein and his seventeen-year-old son, Sher Ali, were both shaking with fear. 'Khuda ki kasam, yahaan koi Pakistani sipahi nahin hai (By God's oath, there is no Pakistani soldier here). They have run away after last light.'

Chhewang's men searched every nook and cranny of the village, but the Pakistanis were truly gone.

The villagers were in a state of shock. Chhewang asked

them to sit down and then spoke in the local Balto language. 'Please do not fear. We have come to help you and free you from Pakistani occupation after twenty-three years.'

Till the night before, India had been a feared enemy, 'a dushman'. The village elders reminisced about life before the Pakistani occupation in 1948 and the adjustments they made under Pakistani rule. However, in the mountains, they felt a stronger kinship with their Ladakhi brethren on both sides of the India–Pakistan border.

In that moment, Ghulam Hussein took a closer look at his guest's face. Chhewang had seemed familiar. Before 1948, Hussein had regularly travelled to the Nubra Valley for trade. He had been to Chhewang's village, and now asked him if he knew Khunzang, his old friend who had two sons. Chhewang, it turned out, was one of those boys.

Marvelling at the coincidence, Chhewang thought about the cyclical nature of life.

History was repeating itself; the women and children of the village were hidden in the nearby nullah, reminiscent of a similar situation decades ago. As a mature veteran of many wars, Chhewang understood their concerns and reassured them of their safety in India. 'The Indian Army will help you . . . I want you to settle down as free citizens of India. India is a democratic country. Here, people of different religions live and work together in peace and cooperation.'

Following this victory, the plan had been to launch attacks on Prahnu and Piun, where the Pakistan Army had retreated. A few hours before the operation, All India Radio announced that the war had ended. India and Pakistan had agreed to a ceasefire at 1700 hours on 17 December. Unlike in 1948, however, Chhewang was satisfied that he had completed his unfinished task. He would go on to receive his second Maha Vir Chakra for his efforts – a rare feat in India's war history.

After the ceasefire, a portion of Baltistan comprising Chalunka, Turtuk, Thyakshi and Thang villages now belonged to India, shifting the Line of Control (LOC) overnight. It was a tactical victory in challenging terrain, and now peace brought a gentler beauty to the countryside.

Chhewang stood in awe, taking in the pristine landscape, reflecting on his journey since volunteering in Leh. Looking up, he thought of his mother for a moment. She would be proud, knowing her son had become Stak, the Tiger.

Postscript

Chhewang retired in 1984 as a colonel and later served as honorary colonel of Ladakh Scouts. He holds the distinction of being the only awardee in the Indian Army decorated during every war he fought while in service.

Major Chhewang Rinchen was awarded the Sena Medal in the 1962 India–Pakistan war.

In Leh, a 400-metre bridge connecting Darbuk with Daulat Beg Oldi is named Colonel Chhewang Rinchen Setu in his honour.

The Tiger of Ladakh passed away in 1997 at the age of 66. The legend of Stak lives on, undefeated.

In 1984, while writing the book *Baltistan Par Ek Nazar*, Muhammad Yusuf Ahidi, a former commander of Gilgit Scouts, who had led forces on the Nubra front in 1948, wrote how the repeated attacks by Pakistani forces in Ladakh had been thwarted by the valour and leadership of a seventeen-year-old boy named Chhewang Rinchen. He writes that 'but for Commander Rinchen foiling our attacks, we would have been the masters of Ladakh'.[30]

5

Rise after the Fall of 1962

The Amazing Comeback of Haripal Kaushik

Haripal Kaushik felt a deep sense of tension in his bones as he watched Raghbir Singh Bhola attempt the backflick. The stadium crowd held their breath as the ball sailed towards the goal but missed its mark, veering wide.[1] Pakistan held on to the goal they had scored in the first half.

It was the end of an era.

On 9 September 1960, in Rome, India's unbeaten streak in world hockey came to an end. Pakistan claimed the gold, ending India's six consecutive Olympic wins that had started from Amsterdam in 1928.

Watching from the sidelines, Haripal, a young forward of the Indian team, was devastated. Four years earlier, in 1956, he had proudly celebrated India's victory in Melbourne, wearing his gold medal and standing on the podium as the Indian flag rose and the national anthem played. Indeed, the pain of losing the gold that year was deeply felt.

Known for his speed and supreme stickwork,[2] Haripal and

his teammates were attacked by the media after the team's return from Rome. *The Hindu* wrote, 'Our defence failed to cope with their pace. Moreover, our forwards were haphazard in their movements and did not do one thing right.'[3]

Haripal's journey to the Indian team was unconventional. As a student in DAV College, Jalandhar, he had been recruited as a substitute by the Sikh Regimental Centre (SRC) hockey team for the All-India Dhyan Chand Hockey tournament in 1955. On that day, playing as the inside left forward, the young boy repeatedly harried the Pakistan defence. SRC won the trophy and Haripal's cool demeanour and skilful artistry had caught the attention of Indian selectors.

While choosing players for the Indian team for the Melbourne Olympics, the selectors realized that Haripal could not be considered as a Services player because he had not been enrolled into the army. Soon after, he was recruited into the 2nd Sikh Regiment as an office clerk, enabling him to officially represent the Services. He secured a spot on the Indian hockey team that won the gold in Melbourne.

By 1959, he had joined the 1st Battalion of the Sikh Regiment (1st Sikh) as a commissioned officer,[4] but his hockey commitments kept him away from service with his battalion.

As the Rome Olympics came to a close, Haripal prepared to rejoin his army battalion, unaware of the storm brewing in the East. Little did he know that he and his country were about to face a tumultuous challenge from the Chinese.

When the Chinese Army occupied Tibet in 1950, India's geopolitical landscape underwent a significant transformation.[5] Sharing a border, India and China became neighbours for the first time in history.

The 1950s marked a period during which the two nations

approached their perceptions of the border differently. In 1954, China challenged the validity of the McMahon Line – a demarcation line proposed by British colonial administrator Sir Henry McMahon in 1914 – which delineated the border between India and China in the east.[6] India, on the other hand, adhered to the Line and downplayed the possibility of a confrontation with China.[7] However, China expanded its boundaries and built a road connecting Tibet with Xinjiang, passing through the Aksai Chin region, a disputed territory. India raised objections, viewing this development as a violation of its sovereignty.[8]

The Tibetan uprising included Lhasa and the Dalai Lama's escape from China to seek asylum in India in 1959 marked the breaking point. That year, People's Liberation Army (PLA) troops attacked an Indian army post at Longju on the North-East Frontier Agency (NEFA, now Arunachal Pradesh/AP) border, followed by an encounter in the Chang Chengmo Valley in Ladakh.

In 1960, while Haripal Kaushik and the Indian hockey team saw their dreams shattered on the hockey field, tensions escalated perilously as China adopted a more rigid stance regarding the McMahon Line, leading to minor border incidents. India's deployment of the 4th Infantry Division,[9] despite being ill-equipped for high-altitude combat in sub-zero temperatures and the challenging NEFA terrain, was made after sidelining General K.S. Thimayya, DSO (Distinguished Service Order), the Indian Army Chief.

Tawang, a prominent town located along the McMahon Line, hosted a significant military base. The 7th Infantry Brigade (7 Brigade) stationed in Tawang was hastily dispatched to occupy defences along the Namka Chu river, facing the

Chinese with little artillery support and no planning.[10] In April 1962, an officer of the 1st Sikh battalion was tasked to set up a post above the confluence of the Rong Chu and Namka Chu rivers. Known as the Dhola Post, it commanded a vantage point overlooking the Namka Chu river.[11]

Defying the basic logic of warfare, the 7 Brigade troops occupied lower ground in their defences and were tasked with confronting significantly larger Chinese forces that held commanding positions on the Thag La ridge. This choice set the stage for a monumental failure, marking the beginning of a series of ill-advised actions that would ultimately result in the disastrous outcome of the 1962 war.

That was a glimpse into the future. For now, let's return to Haripal's story in the present.

In July 1962, coinciding with the escalating border tensions, Haripal had his first posting with his battalion in Tawang. Despite being a commissioned officer, Haripal's army career had primarily revolved around his passion for hockey, and he had never served in the battalion until then. Young officers spent time getting to know their men, learning about their stories, families and villages they came from. An officer was required to know about the skills of the men he commanded, enabling him to assign suitable roles to them during war. While other officers were already familiar with their companies, Haripal was just beginning to understand the unit and the men, all the while coping with the mounting border pressure.

The question loomed: Could a hockey player-turned-rookie military officer lead his men in a crisis?

From sticks to rifles: Embracing a new battle

Located on the Indian side of the Line of Actual Control (LAC), Bum La is a breathtakingly beautiful border pass situated at an altitude of 4,600 metres. Its name, derived from the Tibetan language, signifies 'a pass nestled amidst mountains'. Part of the ancient trade route from Tawang to Tsona Dzong in Tibet, this strategic gateway connects India's Tawang district to the Chinese-occupied Tibet Autonomous Region (TAR) (see map on page 124).

The Chinese had plans to capture Tawang (in AP) and control India's North-East. The quickest way to reach Tawang was via the 26 kilometres track that passed through the Bum La Pass, among the limited motorable routes in the region. Between 10 September and 1 October 1962, 1st Sikh had moved up from Tawang and taken positions on Bum La, tasked with defending the pass against the advancing Chinese forces aiming to capture it.[12]

Lieutenant Haripal Kaushik commanded Delta or D Company, which was stationed at Tongpen La, a mountain pass near Bum La. In the vicinity, around 3 kilometres southwest of Bum La, stood Twin Peaks, at an elevated height providing a vantage point to monitor the Chinese movements. Connecting Twin Peaks and Bum La was the flatland known as IB Ridge. Positioned at IB Ridge – named after an abandoned Inspection Bungalow – was 11 Platoon under the command of Subedar Joginder Singh Sahnan. The rocky escarpment of IB Ridge provided Joginder's platoon with an ideal vantage point to act as a screen,[13] to give early warning to Haripal's Delta Company and engage with the enemy's advancing units to delay their progress.[14] A likely axis of attack on Tawang

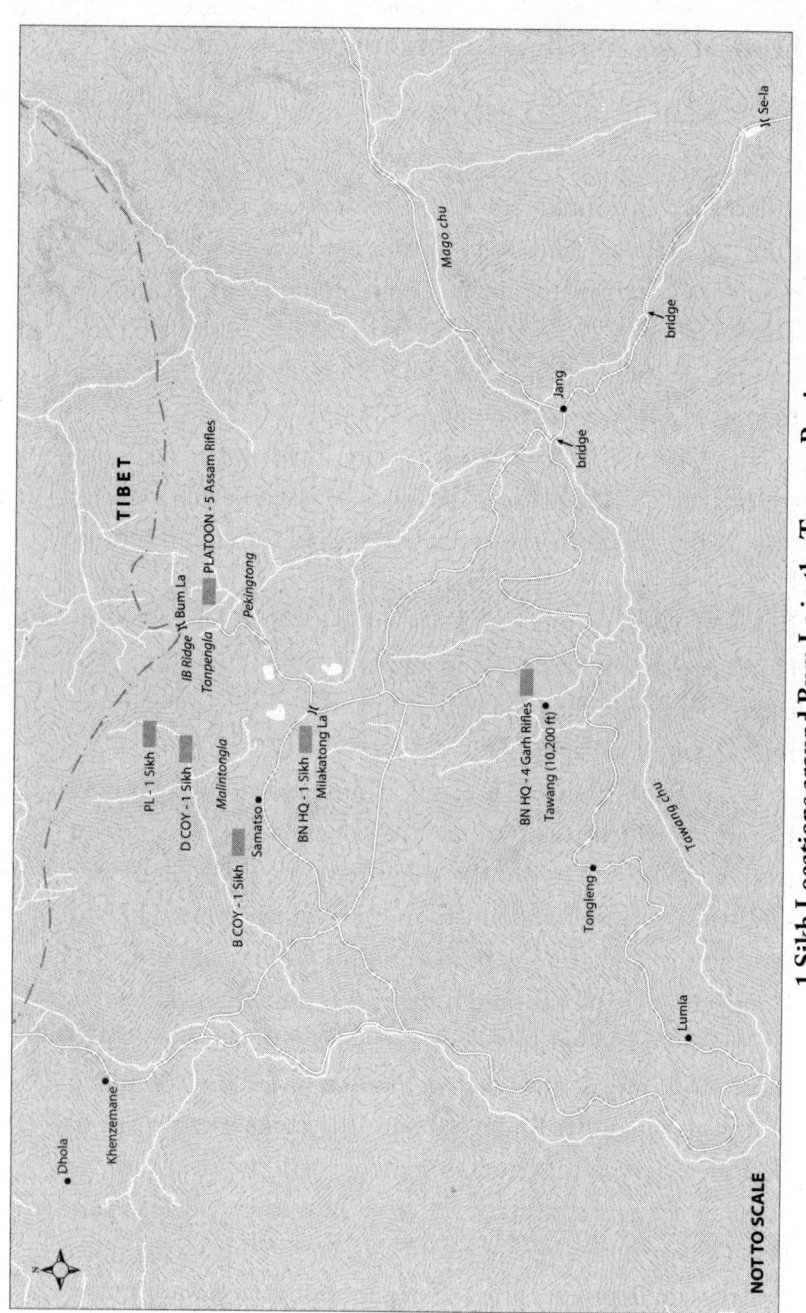

1 Sikh Locations around Bum La in the Tawang Region

was also held by a platoon of 5 Assam Rifles.[15] Different subunits of the Sikh battalion occupied positions along the ridge behind it, as part of their defensive line.[16]

The battalion had five artillery batteries providing support, including three batteries of 97 Field Regiment (25-pound guns) on the road leading to Tawang. Captain Gurcharan Singh Gosal served as the forward observation officer (FOO) for 1st Sikh.[17]

One morning in October 1962, the CO called Haripal to his office. He was told to be prepared as the Chinese were biding their time for the right moment to launch an attack.[18] His hockey days behind him, Haripal was about to enter the realm of active combat. The hockey stick had been replaced by the rifle.

The haunting silence of the imposing Himalayan peaks was a far cry from the electrifying atmosphere of the Olympic stadium in Rome. Haripal pondered the ominous prospect of war in the harsh terrain of the Himalayas.

Bum La loomed ahead, a formidable trial by fire that demanded preparedness. Haripal immersed himself in rigorous preparations. He meticulously studied the terrain, seeking wisdom from seasoned Junior Commissioned Officers (JCOs) like Joginder to ensure his men were well positioned and mentally prepared. He had heart-to-hearts with his soldiers, building trust and fostering cohesion.

Just as on the field, he now applied his tactical acumen to the battlefield. Fauj is similar to sports, he mused, thinking of the triumphs and heartbreaks that followed his Rome low after the Melbourne high in the Olympics earlier. Yet, he understood that in war, the consequences were more profound – the price of lives lost could never be recovered.

While Haripal was siting the defences of his platoons on the ridge, rehearsing the drills for a defensive battle, Chinese PLA forces quietly crossed the McMahon Line and set up four battalions south of the Thag La Ridge, poised to engage in battle along the Namka Chu river.

On 20 October, they launched a relentless bombardment on the Indian positions, marking the beginning of the Sino-Indian war. Simultaneously, the Chinese army also initiated an incursion on the western sector of the border, including the Galwan Valley.[19]

The tactical blunders of the senior army leadership had been exposed in the early stage of deployment in the war.[20] Several senior commanders lacked conviction while dealing with an abrasive Lieutenant General Biji Kaul, 4 Corps commander and a key decision-maker during the war,[21] and an interfering Defence Minister V.K. Krishna Menon. On the other hand, Chinese commanders, who had fought in the Korean War and had a clear objective, launched a massive assault on 7 Brigade.

At 0500 hours, heavy mortar fire rained down on the Indian posts in the Namka Chu region. An hour later, intense shelling targeted the Indian forces at the Dhola Post and the Tsangdhar Ridge along the banks of the Namka Chu. By 0700 hours, the Chinese had completely overrun the Dhola Post, resulting in heavy Indian casualties. Despite valiant resistance from the Rajput and Gorkha Regiments stationed at Dhola and Tsangdhar, they were eventually overwhelmed. Over 282 soldiers from the Rajput Regiment lost their lives defending Dhola.[22]

Within half a day of the battle, the Indian troops suffered such significant losses that they were forced to withdraw across the Namka Chu region. The lack of adequate firepower, insufficient personnel, logistical challenges and poor supply lines further compounded the difficulties faced by the Indian forces.[23] The 7 Brigade was defeated within hours and Brigadier J.P. Dalvi, the brigade commander, became the most senior officer to become prisoner of war.

Events at Namka Chu set a tomcat among the pigeons, completely altering the dynamics of the war. From this point onward, a narrative of withdrawals and subsequent disasters would unfold.

The Chinese wave hits IB Ridge, Tongpen La

Haripal Kaushik and Gurcharan Gosal were following the course of events at Namka Chu. In the Indian Military Academy (IMA), they had trained together. They had familiarized themselves with the area and the axis leading to Bum La. Haripal believed that Joginder would be the right person to man the screen and stop the Chinese advance ahead of the main body of his company troops. He recognized that he was facing a challenge greater than the hockey field in Rome. The Delta Company of 1st Sikh at Tongpen La was getting ready to take on the might of the Chinese.

At daybreak on 23 October, three very lights (flares that illuminate the area around the target) lit up the sky atop the Assam Rifles post at Bum La. The Chinese launched a fierce attack, targeting the stone emplacements known as 'sangars' that were used as defensive positions. One Chinese column directly engaged the defences at Bum La, while another attempted to outflank them. The Assam Rifles post fought valiantly, but they were overwhelmed by the superior numbers of the Chinese forces. The Chinese advanced with astonishing speed, dismantling the confidence of the Indian generals leading the war in the east.

At the IB Ridge, Joginder and his platoon of 30 soldiers were prepared and positioned as a forward protection to the IB Ridge and Twin Peaks. At 0530 hours, the Chinese launched a heavy attack on the Bum La axis, aiming to break through to Tawang. The leading enemy battalion attacked the ridge in three waves, each consisting of about two hundred soldiers.

Joginder's screen and Haripal's Tongpen La defence

Manning the screen, Joginder and his men successfully demolished the first wave,[24] and the enemy was temporarily halted by the heavy losses. Soon enough, a second wave arrived, which they also defended adeptly.[25] However, Joginder lost many of his men in the process.

The enemy had advanced in close proximity to the defences. At one stage, Haripal asked Joginder if he should go 'red over red' – bringing defensive fire over his platoon's position – to inflict more damage on the enemy despite risks to own troops. Joginder replied, 'Haan ji, sahib (Yes sir).' At Gosal's insistence, Haripal approved the use of artillery fire, resulting in significant casualties among the enemy forces.[26]

The courageous Subedar Joginder Singh and his small force managed to repel three enemy attacks, causing 110 enemy casualties, according to sources from the Chinese Military History Society.[27] However, the unequal struggle came to an end as his platoon was overrun, with only three survivors managing to make it back to Indian lines.

In the intense trench warfare, Joginder and his men fought fiercely, displaying bravery and defiance. Regrettably, Joginder was captured. He later succumbed to gangrene from a leg wound after declining treatment.[28]

After overcoming Joginder's 11 Platoon at the IB Ridge, the Chinese forces launched an attack on the defences of the Delta Company at Tongpen La.

Haripal and Gosal had formulated a plan and took control of the situation. The Chinese column was caught off guard by a devastating artillery barrage, splitting them up. Relentless and accurate fire halted the enemy's advance in its tracks. In this manner, Haripal and his men successfully repulsed three successive waves of Chinese attacks.

To deceive the enemy, Haripal cleverly positioned his Delta Company across different points, creating the illusion of a larger defensive force. Leading from the front, the energized leader moved between platoon posts, maintaining contact and motivating his men amid flying bullets and falling shells. Haripal understood the crucial role of temperament in battle and emphasized the value of fire discipline to his troops, stressing the need for controlled and directed fire rather than panicky and indiscriminate shooting during engagement. The valiant Sikhs, supported by accurate artillery fire, inflicted heavy casualties on the enemy, with Chinese historians estimating around 175 killed.[29]

In a surprising turn of events, orders were received by Haripal Kaushik at 1130 hours to execute a tactical withdrawal to the battalion's main defences at Milakatong La, a mountain pass linking Tawang and Tsona City in TAR.[30]

Subsequently, the withdrawal would continue to Sela, another mountain pass serving as a gateway to Tawang, where the 4th Infantry Division aimed to establish its primary stronghold by 26 October.[31]

The Delta Company was tasked to hold Tongpen La until 1500 hours after which Haripal and his men were to act as the rearguard for the main body of the battalion to withdraw to a more distant position.

Perturbed and unconvinced about the decision to retreat, Haripal reluctantly followed the instructions. He understood the difficulties involved in executing a withdrawal while facing a relentless enemy. Such an operation is one of the hardest tactical operations in war, especially in a treacherous terrain like Tongpen La where transporting weapons, equipment and artillery guns through the mountains made it doubly arduous. Restoring the morale of the troops, who were told to withdraw from a battle in which they were doing well, presented another challenge.

As per the withdrawal plan, once the Delta Company reached its destination, the Battalion Headquarters at Milakatong La would commence their withdrawal to Jang, a village located at midpoint between Tawang and Sela. The ultimate responsibility rested on Haripal's shoulders as he safeguarded the battalion's movement by holding Milakatong

La against enemy attacks until the entire battalion had completed a successful withdrawal (see map on page 124).

Despite the continuous Chinese attacks and artillery shelling, Haripal carried out a successful withdrawal in stages, covering the battalion's movement and deftly manoeuvring through difficult terrain.

When soldiers, guns and equipment finally reached Milakatong La, there were no casualties or any damage to the large artillery guns or equipment. The execution of a flawless withdrawal was a tactical victory, guaranteeing the ability to continue the fight from a stronger position.

By 1100 hours 24 October, the entire battalion had concentrated at Sela. It was a significant achievement considering the pursuit by the Chinese forces and ensured the battle's ongoing progression.

With the battle at Sela looming on the horizon, news arrived that Lieutenant Haripal Kaushik had been awarded the Vir Chakra in acknowledgement of his brave defence at Tongpen La and his crucial contribution to the successful withdrawal.

Despite the loss of Subedar Joginder Singh, Haripal's exceptional efforts ensured the safety and preservation of his unit, standing as a testament to his exemplary bravery and selflessness. He had 'handled the withdrawal skilfully'[32] at Tongpen La and Milakatong La, showing courageous leadership that gave a significant morale boost to the entire unit.

Moreover, while the decision to withdraw led to significant losses of guns for artillery batteries elsewhere, Gosal and Haripal were the only exceptions, as the guns of 97 Battery – which gave good support to 1st Sikh at Tongpen La[33] – returned undamaged.

Journey to Sela and the collapse of a fortress

For over two years, the government in New Delhi had prepared Tawang as a stronghold against a Chinese attack. However, in adversity, they chose to abandon Tawang and fight the battle at the Sela massif instead.

The sudden withdrawal of the battalions to Sela, a formidable ridgeline running from the border with Bhutan on the west to the Kameng river in the east, was a hastily taken decision due to the escalating situation.

Indian battalion and company commanders on the ground were, however, determined to defend their positions against the advancing Chinese forces. In November, the battle had shifted to the defensive perimeter created by the Sela–Dirang and Dzong–Bomdila areas.

By 24 October, the tactical headquarters of the 4th Infantry Division had been set up at Dirang Dzong, 64 kilometres southeast of Sela. The 62 Brigade, which had been dispatched to provide reinforcement to Tawang, had received new orders to hold Sela. Brigadier Hoshiar Singh had assumed command of the brigade.

By the second week of November, the 62 Brigade had established itself at Sela.[34] At that point, the brigade commanded a total of five infantry battalions. Among them, three battalions – 1st Sikh, 2 Sikh LI, and 4 Sikh LI – were responsible for securing positions on Sela or the surrounding features near the pass. Additionally, the 4 Garhwal Rifles held the screen position northwest of Sela.[35]

Hoshiar Singh, an experienced military leader decorated with the Croix de Guerre in WWII (a prestigious French military decoration), reiterated his resolve to defend at Sela. Extensive patrolling was conducted to gather information on the potential enemy approach routes.[36]

Referred to as an impregnable fortress, Sela was expected to hold off the fiercest waves of attacks. However, the earlier decision to withdraw to Sela had been made hastily, resulting in battalions having to occupy defences without adequate preparation. 1st Sikh (Haripal's battalion) was given a wide frontage to defend but most days were spent preparing field defences and conducting patrols. Essential weapons and stores such as mines, wires, sandbags or medium machine guns were unavailable. As temperatures plummeted below freezing levels, the harsh conditions posed a new set of challenges. The battalion now faced two formidable adversaries: the enemy and the weather.

Having prepared three divisions for the attack, the Chinese Army realized that they could outflank the Indian defences at Sela and strike the divisional headquarters at Dirang Dzong.[37]

While the attack on Sela was carried out by one division, two Chinese divisions bypassed Sela to hit Indian defences in the rear.

The Chinese initiated their assault on Sela by approaching from the north at approximately 0500 hours on 17 November.[38] A plan was hatched to send a column to Nyukmadong, where they waited near a bridge for retreating Indians to arrive from Sela.[39]

The continual retreat from their positions made the Indian moves predictable, while the Chinese plans, remarkably clear, had a sinister brilliance.

In response to the Chinese attack, frantic orders arrived for the brigade to withdraw from Sela on 18/19 November. Hoshiar Singh's objective was to maintain Sela as a stronghold, but Major General A.S. Pathania, the new commander of the 4th Division, had other plans. Taken aback hearing that the enemy had seized nearby areas like Poshing La and Thembang, and threatened the Dirang Dzong–Bomdila road,[40] he instructed Hoshiar Singh to withdraw troops from Sela to Dirang Dzong.

Even though the ground commanders had demonstrated their capability to fight and protect, the determination of the leadership to defend a stronghold crumbled in response to each Chinese threat. By that time, haunting memories of the Namka Chu incident had intensified and loomed large in their minds.

Despite being bound by orders, Hoshiar Singh fought the tactical battle well and his battalions beat back several Chinese attacks on 17 November.[41] Chinese infiltrating parties launched sporadic attacks on the 1st Sikh at Sela, wedging themselves between Haripal's Delta Company and the Bravo Company, but the two companies held the Chinese at bay.

However, the night's battle, coupled with disrupted communication channels, led to confusion the following morning as the brigade commenced its withdrawal on 18 November.[42] The headquarters of the 62 Infantry brigade, three battalions and administrative elements were headed along a road to Senge Dzong, held by an Indian army battalion (13 Dogra). On 18 November, 1st Sikh joined the withdrawing vehicular column of several battalions.[43] The Sela garrison was walking into a waiting ambush.

At the bridge in Nyukmadong, the waiting Chinese ambush party surrounded the column and brought down a heavy volume of fire.

Vehicles were destroyed and Indian soldiers, caught in a well-crafted ambush, either escaped or were killed.

Those who managed to flee headed towards the hills, spreading out in the vast emptiness, hoping to find a path the Chinese hadn't discovered yet.

The war was over for the Indian soldiers. Having accepted their defeat, there was now a desperation to survive, with every man looking out for himself.

A perilous escape

To evade their attackers, the 1st Sikh unit separated from the rest of the column. Tragically, many of their soldiers, including their commanding officer Lieutenant Colonel B.N. Mehta, were killed in the ambush.

Haripal and 14 other soldiers managed to break through the enemy attack. With no knowledge of what had happened to the rest of their unit, they desperately sought an escape route. However, all paths were blocked by Chinese patrols actively searching for Indian soldiers. Meanwhile, the Chinese forces were feeling triumphant after their successful attack, which claimed many lives, including that of Hoshiar Singh.

Haripal's escape from the ambush was only the beginning of his trials as another equally challenging ordeal awaited him.

Moving cautiously through the mountains, deliberately avoiding the expected paths, the group of survivors from 1st

Sikh made every effort to keep their morale high. They had been reduced to a dishevelled group of individuals, relying on the land for survival, sleeping wherever nightfall caught them and taking turns to keep watch while the others rested. With their rations depleted, food became scarce.

Haripal sustained himself on meagre amounts of water found in the streams, doing his best to fill his belly. He felt a sense of duty towards his fellow survivors, as he was still their leader, and it was up to him to keep their spirits up. Memories of the recent past raced through his mind. He couldn't help but think that if they had held the enemy longer, it could have given their forces time to regroup and fight back, preventing the capture of Bum La and then Sela. Instead, they now found themselves struggling in the mountains, trying to lead a small group to safety.

He thought about how the war had turned into a series of withdrawals. He had believed his unit could withstand the Chinese forces, but orders to pull back had disrupted their plans. Now, they were evading an enemy they had recently faced head-on. Since then, they had been continually retreating.

Such thoughts would come and go, but Haripal was not the type to harbour grudges. Instead, he remained focused on finding a way back to safety.

Hope began to fade as their health started to deteriorate. Dysentery and diarrhoea plagued the men, while sores from wet socks worsened, forming painful blisters on the soles of their feet.

One day, Haripal sat beneath a tree, quietly observing the serene waters of a stream below. His clothes were worn and torn, his hair appeared dishevelled, and he sported a rugged stubble on his face.

Three years earlier, during his commissioning into the 1st Sikh Battalion at the IMA, Lieutenant Kishen Khorana, his senior and adjutant in the battalion, had shared the story of how they had pulled strings to secure Haripal's assignment. Allowing himself a brief smile, Haripal couldn't help but wonder about their thoughts concerning him now. Were they still alive? Had they resigned themselves to his presumed death?

The 1st Sikh had experienced a significant setback. The CO and more than 170 officers and soldiers had either lost their lives or were unaccounted for. Some survivors had managed to take a route through Bhutan and reach the battalion base in Bomdila, utterly fatigued and exhausted from the arduous journey across the mountains. Haripal Kaushik, the hero of Tongpen La, was nowhere to be found. Two weeks passed, yet there was no news of his whereabouts. Although the battalion clung to hope, they understood that the chances of ever seeing him again were slim.

Then, one day, a group of survivors arrived at the new battalion base in Charduar, near Tezpur, Assam. Haripal and his men had made it back home.

They appeared as mere shadows of their former selves, all skin and bones. Among them, Haripal was in the worst state. For him, the war had stretched on longer than most. Being the first to engage the Chinese forces at Tongpen La, he had fought valiantly and executed a remarkable retreat in the most difficult circumstances. He and his Delta Company had been a source of pride for the Indian Army.

However, the events following the withdrawal from Sela had traumatized him. Noting his condition, Haripal was immediately evacuated to a military hospital for treatment. The new CO of the battalion granted him a two-month leave, understanding the need for him to recover and heal from the ordeal.

After his sabbatical following the war, Haripal returned to duty but the anguish of losing comrades in battle had begun to affect his well-being. He gained weight and appeared unfit.

Haripal had received the Vir Chakra for his remarkable bravery, but he believed that the senior army leadership had failed the courageous officers and soldiers on that count.

A mentor and antidote

Lieutenant Colonel Karnail Singh Sidhu had been appointed as the new commanding officer of the 1st Sikh Battalion. An astute judge of officers and troops, Sidhu immediately set to work, determined to rebuild the battered battalion and restore their morale by instilling self-belief and mental strength.

Haripal became his primary focus. Sidhu believed that the best way to help him rediscover his inner strength, and give him a renewed sense of purpose, was to revive Haripal's love for hockey.

The hockey fraternity, however, wasn't always kind. There were taunting remarks about Haripal's thwarted hockey career as his fitness – once the cornerstone of his game – was gone. With smirks, people would say that there was no juice left in him now. Noticing that Haripal appeared disheartened, Sidhu took him aside and encouraged him to train. 'I want you to play like you did,' Sidhu told the diffident young war hero. He was put on a simple diet of dal–roti and a fitness regimen was introduced.

Sidhu was a good judge of character and knew Haripal had it in him to succeed again.

Haripal would watch the young battalion boys playing hockey on the cantonment grounds.

They transported him back to his village days in Khusropur, Kapurthala district of Punjab, where his fondest memories of boyhood were intertwined with hockey. The village life revolved around the art of dribbling, weaving and dodging the hockey ball with skilled stickwork.

An inspired Haripal began practising and soon found his old rhythm. Within a few months, with Sidhu's encouragement, he stepped on to the ground. In the infantry, the families of officers and men play a crucial role. In Haripal's case, the officers and ladies of the unit were wholeheartedly involved in cheering him on at every occasion on the ground. The entire unit now wanted their hero to succeed.

At this time, the second half of 1963, preparations were under way for the 1964 Olympics in Tokyo. The Indian Hockey Federation announced a list of probables and to everyone's surprise, Captain Haripal Kaushik had been called up to join the training camp.

However, Haripal harboured uncertainty about returning to play for India. Some hockey experts expressed concerns about his fitness, questioning whether he could compete at a high level, let alone participate in the Olympics.

Yet, as Sidhu, his fellow officers and the ladies of the battalion cheered him on, Haripal had a flashback to a hockey match in 1955 between an SRC team and the Aryan Club of Lahore. In that match, a remarkable young boy from

DAV College showcased his exceptional speed and ability to outmanoeuvre the Pakistani defence repeatedly. One of the selectors closely observed the match, taking note of the name of this talented young player: *Haripal Kaushik*.

Haripal decided to give the Olympics a shot.

At the three-day trials camp in Jalandhar, forty players battled it out for a spot on the eighteen-member hockey squad for the Tokyo Olympics. Haripal made the list. The veteran hockey star and war hero was back.

Haripal's last match was at the Olympics in Rome. In 1962, while Haripal was at the front, the Indian team competed in the Asian Games but once again lost to Pakistan. Determined to reclaim the top position and win the gold, the Indian Hockey Federation dispatched the team on a preparatory tour to New Zealand. Haripal, who had been part of two previous Olympic squads, was now considered a senior team member.

His two appearances for the nation were tragically separated by a devastating war, whose repercussions were keenly felt. While on tour in New Zealand, players would often get together and sing patriotic songs. Ramesh Saigal's 1948 film *Shaheed* (The Martyr) had captured imagination. Before the first match, legendary goalkeeper Shankar Laxman had initiated a sing-along of 'Watan ki raah mein, watan ke naujawan shaheed ho (For the sake of your nation, oh youth, sacrifice yourself).'[44]

Newer talent had emerged, and among them was a promising youngster named Harbinder Singh who ran like the proverbial wind, leaving defences in his wake and outmanoeuvring defenders before they could even prepare to tackle him. In New Zealand, Haripal formed a formidable partnership with him, combining their skills and abilities on the field. The team had a strong attacking line-up with numerous skilled forwards, and it was encouraging that Haripal had found his form quickly.

The team suffered a loss in their first match but made a resounding comeback with a 5–2 victory in the second. In the third match, they tore apart the Kiwi defences, triumphing with a commanding 8–2 scoreline. The local newspapers praised the destructive Indian forward line, expressing their opinion that this Indian team would prove challenging to defeat.

A crucial moment during the third match in Wellington spoke volumes about Haripal's form.

Faced with two defenders marking him closely, he made a daring decision to take on one of them. With remarkable

Haripal Kaushik dribbling past a defender

speed, he swiftly moved into open space, dragging the defender along while evading the frontal block attempted by the second defender. Without compromising his pace, Haripal skilfully controlled the ball, feinting past the first defender and causing him to move in the wrong direction. Then, with a sudden change in direction, he left the trailing defender in a state of confusion, rendering him ineffective.

The two defenders went hard at Haripal but instead collided with each other, while Haripal sped ahead – unmarked and into an open space – all the while not losing the ball for a moment and set up a goal.[45] Young Harbinder Singh, whose blazing speed often left defenders breathless, was awestruck by this extraordinary display of dodging, feinting and explosive acceleration.

Crossing swords with bête noire

Amid the mounting pressure of expectation, the Indian team arrived in Tokyo for the Olympic Games. They had a poor start, drawing two crucial matches out of the first three.

The team's vaunted forward line had not delivered the goods. India was in imminent danger of elimination in the group stages.

Another unforeseen challenge arose. Indian players were playing in Bata shoes without studs in the soles. The rain during the tournament had rendered the playing surface slippery and wet. Without proper shoes, the players kept slipping frequently, affecting their performance. It rained heavily before the match against Canada, causing concern. As a solution, the team was given football shoes to play hockey in. But the leather on the shoes expanded when they became wet.

As a makeshift solution, after the matches, the players dried their shoes using the heater in their hotel room. When word got around, Adidas provided them an extra pair of shoes.[46] This helped, and India rolled past Canada, but the formidable Dutch awaited them.

Their backs against the wall, India stormed into the next match, and outplayed the Dutch. They had found their rhythm and went on to win the rest of the group matches. In the semifinals, they came in from behind to trounce Australia 3–1.

India and Pakistan were indeed set to cross swords once again.

On the morning of the finals, Mohinder Lal, an Indian midfielder, shared a private conversation with Harbinder Singh.[47]

'I have been thinking about this . . . today, we will get a penalty stroke, and I will hit the ball at the roof of the goal post.'

'Why?' Harbinder asked, curious.

'Their goalkeeper is short, he won't be able to reach for it,' Lal said with a smile.

Pakistan had entered the finals in impressive form, having won all their previous matches. Meanwhile, India had the triumphant return of Haripal Kaushik, whose performance throughout the tournament had been remarkable.

For Haripal, it was a journey coming full circle, rebounding from the aftermath of a major war and now on the verge of achieving Olympic hockey glory.

On the cold afternoon of 23 November, the Komazawa

Olympic Park Stadium in Tokyo was packed with people eager to watch the giants of hockey facing off again. Relations between the two teams were not cordial, and the players refused to communicate.[48] An umpiring incident triggered a fracas, leading to a temporary pause in the match for the players' tempers to cool down.

The game had opened at a scorching pace, but no goal was scored in the first half. Haripal and the other forwards maintained relentless pressure on the Pakistani defence. Five minutes into the second half, India was awarded a penalty corner. Penalty corner specialist Prithipal Singh drove a fierce hit from the top of the striking circle. His shot struck Pakistani captain Manzoor Hussain's foot on the goal line, handing India a penalty stroke. Mohinder Lal stepped up to take it. He took a quick look at the Pakistani goalkeeper and flicked the ball to the top of the goalpost. India went up 1–0.

His conversation that morning had indeed been prophetic!

Stung by the goal, the Pakistanis came back hard, winning penalty corners and making forays into the Indian defensive circle. Indian goalkeeper Shankar Laxman thwarted every attempt that came his way. In response, Haripal, Harbinder and the forward line launched counterattacks on the Pakistani goal, exploiting the gaps. Right before the final whistle, Pakistan won a crucial penalty corner. The crowd now waited with bated breath.

The Indian defenders successfully cleared the ball just as the final whistle blew, signalling the end of the match. The crowd erupted and the Indian players on the ground broke into an impromptu bhangra. India's hockey pride had been restored.

Fans swarmed the streets in cities across India, distributing sweets. Newspapers were busy printing paeans for a hero who had been written off. As radio commentators enthusiastically broadcasted the news to millions in India, Haripal's unit revelled in joy, sharing sweets to celebrate. Amid the joyous soldiers and officers in the Sikh paltan, a beaming Lt Colonel Karnail Singh Sidhu basked in contentment, knowing that his mission had been accomplished.

Haripal had brought home the gold!

But more importantly, he had found inner peace. He had fought until the end, just as he had hoped to do with his troops in the mountains before they were abruptly and repeatedly called back.

Haripal had found closure in the game he loved.

Colonel Haripal Kaushik

Haripal Kaushik's Olympic gold

Postscript

In the 1966 Asian Games held in Jakarta, India and Pakistan faced off in the final. Once again, India emerged victorious, avenging their previous defeat in the Games. Haripal, now vice captain of the Indian hockey team, had reclaimed both the Olympic and Asian Games' gold medals.

Awarded the Vir Chakra for gallantry, Kaushik won two Olympic gold medals, one silver, one Asian Games gold medal and the Arjuna Award for sports, etching his name in Indian military history for excelling in both the sporting and military realms. He retired on his own terms, as he desired. He left the sport at the pinnacle of his career, shining brightly and at the top of his game.

6

Top Guns of Boyra

Gentlemen at War

That morning in 1997, Donald Lazarus was surprised to hear the doorbell ring. Visitors at the front door were a rarity in Coonoor, a serene hill town. Accustomed to the gentle chipping calls of the Nilgiri Flycatchers outside his window, he cautiously opened the door to find a postman standing there with a letter for him.

Lazarus was a familiar face in Coonoor, having been living there for five years and working for the Christian Mission Service. He led a quiet life, and at times he found it hard to believe that not long ago he had been on track for a high-ranking position in the Indian Air Force (IAF). However, he made the choice to leave the service and settle down in the hills.

A charming Siberian Rubythroat, on its migratory journey from a distant place and time, unexpectedly graced the window. Another surprise, Lazarus mused. He looked down at the envelope in his hand once again, then quietly put it away.

The letter, from the Pakistan Air Chief, had unleashed a flood of memories.

A prelude to war

In 1971, Pakistan was in a state of turmoil. Sheikh Mujibur Rahman and his Awami League had won an absolute majority in the National Assembly elections held in December 1970, sweeping nearly all the seats in East Pakistan (now Bangladesh). They now had the mandate to form the government in Pakistan. However, the West Pakistan establishment, led by President General Yahya Khan, refused to transfer power. On 1 March 1971, Khan suspended the National Assembly's inaugural session.[1]

Upon hearing this, widespread protests erupted in East Pakistan.[2] In response, Khan resorted to deploy the Pakistan Army to suppress the non-cooperation movement. This marked the beginning of the savage Operation Searchlight, during which protesters and sympathizers of the Bengali cause were brutally massacred.[3] On 25 March, Mujibur Rahman was taken into custody by West Pakistani soldiers.[4]

India's Prime Minister Indira Gandhi visited several countries in October and November of that year, pushing for sanctions against Pakistan. In response, on 18 November, French President Georges Pompidou, among others, wrote a letter to President Yahya Khan urging him to release Sheikh Mujibur Rahman from custody as part of a political settlement.[5] This was rejected by Khan. As diplomatic efforts reached an impasse, the situation on the ground grew increasingly precarious, with millions of refugees flooding into India from East Pakistan, placing immense pressure on India's eastern states.

Tensions and minor clashes prevailed along the India–East Pakistan border. It seemed that the two nations were on the path to another war, merely six years after the 1965 conflict. Preparations and contingencies for such a scenario were under way.

The bugle of war rang out in the sky, although the first signs of it appeared on the ground.

The Boyra salient is a finger-shaped land protrusion that extended from India into East Pakistan. Located approximately 100 kilometres northeast of Calcutta (now Kolkata),[6] the geographical separation was due to the border alignment.

The hamlet of Garibpur, located 9 kilometres inside the Boyra salient and northwest of East Pakistan, is situated along the highway from India to Jessore (East Pakistan). Due to its strategic location, the salient was an important crossroad for both the nations.[7]

Following Operation Searchlight, tensions along the Indo-Pakistan border escalated. The Mukti Bahini, a local guerrilla resistance force backed by Indian forces, conducted military operations, prompting the Pakistan Army to infiltrate into Indian territory, resulting in casualties on the Indian side. In response to the escalating situation, the 14th Punjab Regiment of the Indian Army was moved to Boyra to secure and control the border.[8]

To protect their position, the Indian forces decided to prepare for an offensive operation by capturing Garibpur. Late at night on 20 November 1971, after a harrowing crossing of the Kabadak river,[9] the infantry battalion of 14th Punjab along with C Squadron, 45 Cavalry equipped with 14 PT-76 tanks (Soviet amphibious light tanks), established a foothold in Garibpur. These units were part of the 350th Infantry Brigade, working in conjunction with the 42nd Infantry Brigade. By early morning, the battalion had gotten into position, and the men were working hard to finish preparing the defences before sunrise.

At dawn, there was a fierce response from Pakistan's 107 Infantry Brigade. The 3 Independent Armoured Squadron of Chaffee tank destroyers launched a counterattack against the advancing Indian troops.

Lieutenant Ajit Apte, a young artillery officer, was deployed with 99 Field Battery (Q Battery), providing crucial support[10] to India's 45 Cavalry regiment. At dawn on 21 November, Apte watched from the flanks of the artillery regiment as Pakistan launched their counterattack. A fierce clash of tanks ensued.

Positioned on an anti-tank platform, Apte's guns were loaded and prepared to engage as the enemy tanks advanced in front of his artillery position. He was about to fire when Naib Subedar Raghunath Singh, a veteran of the Indo-Pak war six years ago, stopped him. 'Sahib, don't give fire orders. It seems the enemy has not seen us. If we open fire too early, they may target our battery and the surprise would be lost,' he whispered. Trusting his judgement, Apte held his fire.[11] As soon as they had the enemy in their sights, India's PT-76 tanks opened up, helped by the artillery battery.

Thirteen Pakistani tanks were destroyed that day, while India lost four.[12] Apte learnt an important lesson in the value of fire discipline and deft timing, especially when the enemy was momentarily at an advantage, which would prove helpful a day later.[13]

The Pakistanis may have suffered a defeat on the ground, but it was far from over. The fight was about to go airborne (see map on page 152).

Battle takes to the skies

The quick reversals in the tank battle had upset the Pakistan Army commander, who called for air support[14] (see map on page 152).

In the east, the Pakistan Air Force (PAF) had a squadron of F-86 Sabre fighter jets at the Tejgaon airfield in Dacca (now Dhaka). Among the fighter pilots listed for the mission was Flight Lieutenant Parvaiz Mehdi Qureshi, a daring pilot with an impressive track record, having earned the prestigious Sword of Honour – awarded to top-ranking cadets training in the Pakistan armed forces academy. (In the same batch was his friend and namesake, Pervez Musharraf.)

Qureshi, a young pilot, had his abilities put to the test during the first Indo-Pak war in 1965 and he was familiar with Indian tactics of air combat. In 1969, he was fast-tracked into commanding 14 Squadron – the PAF's first fighter-bomber squadron – and was stationed in East Pakistan.

The 14 Squadron, known by the nickname 'Tail Choppers', was the Pakistani unit responsible for defending East Pakistan airspace against the larger Indian forces. Anticipating an impending war, Pakistan initiated efforts to strengthen its air force. Sixteen Canadian Sabre Mk. 6s

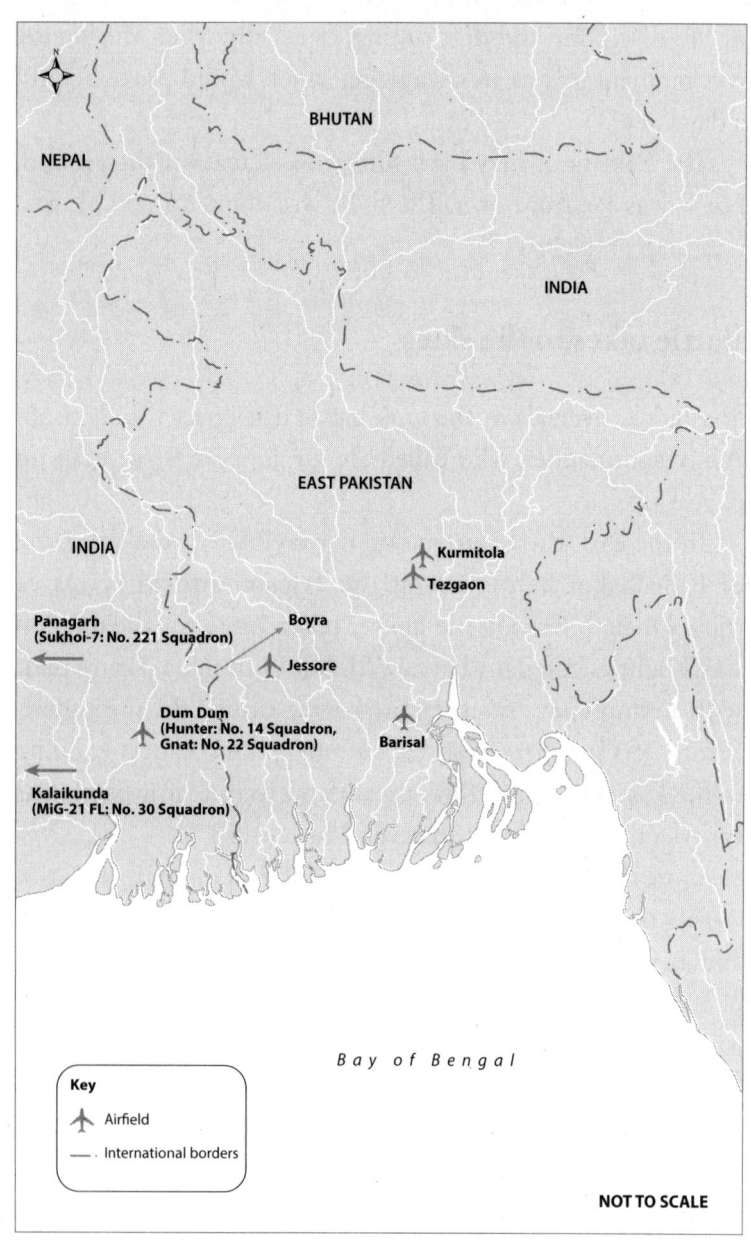

Map of East Pakistan, 1971

were covertly transported through Iran following a deal with West Germany. These aircraft, Canadian variants of the F-86 Sabre, were upgraded with AIM-9 Sidewinder missiles (a short-range air-to-air missile) and the utilization of the more potent Avro Canada Orenda (jet) engine.

The task of safeguarding the air defence of the Calcutta sector was assigned to India's 22 Squadron, stationed in Kalaikunda. The IAF activated a detachment from Dum Dum airfield in Calcutta. The unit was helmed by Wing Commander B.S. Sikand, who had been a POW in Pakistan in 1965.

Flying Officer P.K. Tayal, one of the Indian pilots, had borrowed a colleague's G-suit for a hastily organized mission the previous day. The G-suit, a specialized garment worn by pilots to counteract the effects of high gravitational forces during flight manoeuvres, was a smaller size. Tayal was uncomfortable during the mission, so he spoke with Donald 'Don' Lazarus, also stationed at the Kalaikunda airbase, and asked Don to send his G-suit along with spare parts that were being sent to Dum Dum.

Don couldn't locate Tayal's G-suit, so he flew down to Dum Dum, knowing that a detachment would be stationed there soon due to the events of the previous day.

Tayal was not pleased about relinquishing his position to Don. He flew back to Kalaikunda in a replacement Gnat brought by Tayal's coursemate, Sunith 'Su' Soares.

Tayal's missing G-suit would have an impact on the destiny

of the flyers. Little did Don Lazarus and Su Soares know what they had set themselves up for![15]

On the morning of 22 November, at 2000 hours, Indian radars picked up the whirring of four F-86 Sabres in the Jessore area. India's 22 Squadron was in readiness at Dum Dum.

A scramble was sounded and four pocket-sized Folland Gnats (jet aircraft)[16] soared into the skies from Dum Dum. Led by Sikand, the formation had Don Lazarus as his No. 2, with second section pilots Flight Lieutenants Roy 'Mouse' Massey and Gana 'Gun' Ganapathy.

Meanwhile, four F-86 Sabre fighter jets of the PAF strafed India's ground forces of the 305th Infantry Brigade. (Strafing refers to an aerial assault where fighter jets fly at low altitudes and release a continuous stream of gunfire on to ground targets.)

From his concealed bunker, Apte saw one of the Sabres head down, straight for the bunker.

A flurry of bombs rained down on the ground around him, with a shell landing 25 metres away from the command post. No one was injured.

The objective of the mission was to locate the Indian gun positions and strafe them. Two more Sabres were waiting for the guns to take the bait and reveal the positions.

The plane was approximately 100 yards away, and Apte had a clear view inside the cockpit. He pondered whether to open fire on the aircraft, but suddenly, he remembered Raghunath Singh's advice, which held him back from taking action.

The Gnats appeared on the horizon just as the Pakistani Sabres flew away.

Squadron Leader Dilawar Hussain had confidently led the PAF mission, growing in stature with each unopposed flight over Indian defences. They continuously bombarded the Indian ground troops at Garibpur without facing any obstructions.

When Indian radars detected the sound of four F-86 Sabres once again, 22 Squadron at Dum Dum airfield was ready for takeoff with Sikand leading the formation comprising Don as his wingman (No. 2), and Mouse and Gun in the following section. For the second time, the Indian formation returned without achieving its objective.

Flying Officer M.A. 'Gun' Ganapathy came back an unhappy man that day. He had been stationed at Dum Dum airfield in Calcutta for several months. During peaceful evenings, he often pondered whether the eastern front would ever see action. Previous conflicts with Pakistan had always taken place in the west, and he dreaded the idea of being a mere spectator to a distant war. The young pilot couldn't help but wonder, 'Our duty is to defend Calcutta's airspace ... Who would dare attack a city?' These questions continued to trouble him.

On the previous mission at Garibpur, Gun was the first to spot a Sabre in the skies. He radioed Sikand, who didn't respond. (Sikand later clarified that he had not received the call, indicating there was some miscommunication.) Back at the base in Dum Dum, Gun was frustrated that the lack of communication between him and Sikand had caused them to miss their one chance at the Pakistanis. They decided to rearrange the line-up. Sikand, having flown twice that day and feeling tired, chose to step down. Taking his place was Su Soares, the pilot who had arrived at Dum Dum from Kalaikunda the day before.¹⁷

The forward line of the 4th Battalion, Sikh Regiment (4 Sikh) had also spotted the Sabres.

Flying Officer S.V. Savur, a Forward Air Controller (FAC), had desperately tried to direct the Gnats to the Sabres using his very high frequency (VHF) set. The FAC is an air force pilot who accompanies troops on the ground and establishes communication, via a VHF radio communication system, with airborne colleagues providing support to the troops.

The Sabres faced no resistance from the Gnats as they were not intercepted.

Under the relentless assault of aerial attacks and tank fire, the enraged soldiers of the Sikh battalion and artillery unit vented their anger with a string of verbal insults.

When would the Gnats make their move?

Third-time lucky

The Pakistani sorties were turning more audacious with each successive mission. Despite the setback in Garibpur, where Indian troops had captured ground earlier, the Pakistani

Sabres had dominated the skies, outsmarted Indian Gnats and bombed targets at will. It was a matter of time before they made a dent in the Indian defences.

The sun was still up, and the pilots were eager. A third strike was planned by the confident Pakistanis that day. Wing Commander Afzal Chaudhary would lead a team of Squadron Leader Dilawar Hussain, the subsection leader and Flying Officers Sajjad Noor and Khalil Ahmed as wingmen.

Moments before they were set to take off, a dejected Parvaiz Qureshi saw his name missing from the list. He approached Hussain and expressed his disappointment at being left out of the mission. Impressed by his enthusiasm, Hussain said, 'You're on!' and assigned Qureshi his position.[18]

That afternoon, at 1420 hours, four Sabres took to the skies. After take-off, Sajjad Noor's plane developed a radio communication system malfunction and had to return to base.

The Pakistani raid now had three pilots instead of four. The downside of such confidence was that the forays had become too predictable. Savur recalled how they could set their watches for the first strike at 0900 hours, followed by another at noon, and then a third one three hours later. 'They (the enemy planes) would dive and strafe in turn in an unhurried manner,' he recalled.[19]

At Dum Dum, the 22 Squadron pilots were awaiting the arrival of the enemy.

Flying Officer K.B. Bagchi, the Fighter Controller, called to test the dedicated landline.

Gun answered. 'Put us on Standby 2 if you like,' he urged. Su and Don, who were listening, privately hoped Bagchi would stick to Standby 5. (Standby 5 meant pilots would take off within five minutes, while Standby 2 required them to be ready in just two minutes.)

On the tarmac, despite the November winter, the sun was at its brightest.

'Sure, sir,' Bagchi's voice boomed on the phone, confirming Gun's instructions on upgrading to Standby 2.

Inside the tent, the radio blared songs and, intermittently, the news. Gun was sceptical about going another round. '. . . little hope this afternoon . . . two sweeps by PAF already . . .' he muttered.

Lunch over, Gun turned the pedestal fan towards himself and sat down on his camp bed to write a letter home. Don and Su were playing Scrabble, and Mouse was busy reading. At 1440 hours, the klaxon – a loud electric horn – went off. Bagchi's familiar voice rang over the PA system for the third time that day.

'*Scramble, scramble, scramble!*'

The four Indian pilots dashed to the aircraft, jumped into the cockpits and strapped themselves swiftly, saving a minute in time. The Gnat engine roared to life and was ready to take off by the time the pilots had secured their helmets and closed the canopy. Before the planes hit the runway, Dum Dum ATC radioed: 'Cocktail formation, cleared for take-off, all four aircraft.'

No words were exchanged. They took off in pairs, retracted

their landing gear and flaps and swiftly flew off, propelled by the north-easterly wind. Mouse had deduced that the Gnats missed their mark earlier because the leader hadn't flown in full throttle. In the lead now, he pushed the lever fully forward. They rapidly climbed to an altitude of 1,000 feet, soaring at a speed of 500 knots (900 kilometres/hour), racing over villages along the highway.[20]

The distance to Garibpur was about 80 kilometres, roughly eight minutes of flying time in the Gnat.

Su, positioned slightly behind Mouse, ensured he remained within his line of sight. From the ground, Bagchi made strenuous efforts to provide them with directions over the VHF system. However, the combination of low altitude, whistling wind and cockpit noise posed challenges to effective communication.

But the boys from Boyra were determined to face the sophisticated Sabres over Garibpur. The skies were now set for a dogfight.

Meanwhile, after shelling a tank formation on the ground, the roving PAF Sabres were looking for more targets over the Indian position. An artillery unit that had targeted them caught their attention and soon Chaudhary was leading an attack on the battery.

On the ground, a large group of soldiers and officers awaited the upcoming action in the sky.

Captain Harkirat Singh Panag of the 4th Battalion, Sikh Regiment (4 Sikh)[21] was the adjutant of his unit who ran its combat operations.[22]

On 22 November, at 1500 hours, Panag was returning to the unit headquarters from the logistics base when he saw three Pakistani Sabres coming in for the last sortie (mission) before sunset. 'The Sabres were carrying out high dive attack on our positions. In turn, the Sabres were climbing up to 2,000 feet and coming down to 500 feet for weapon release like the German Stuka bombers. Our medium machine gun and light machine guns were engaging the aircraft.'[23]

Suddenly, he saw four fighter aircraft fly in from the east. As the fighter planes flew overhead, Panag felt the ground shake and his jeep rocked violently. 'Seconds later, three aircraft from the second mission peeled out of formation and headed for the Sabres, which, oblivious to their presence, were continuing with the dive attacks. It was clear that our Gnats had joined the battle. I stopped the jeep and stood waiting for the 'dogfight' to begin,' he recalled.[24]

Mouse ascended to an altitude of 3,000 feet, leaving the noises behind as they faded with increased altitude. Flight controller Bagchi provided updates on the radio, directing the pilots, 'Six kilometres, 2 o'clock.' Mouse scanned the area ahead, searching for any signs of the enemy.

Su occupied the outermost position on the left side of the formation, with Mouse on his right. Don flew on the

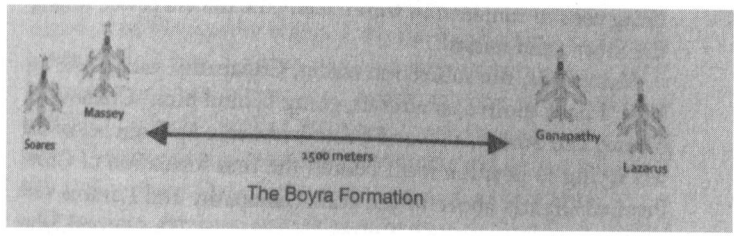

The formation

Source: S. Jagan Mohan, P.V., and Samir Chopra, *Eagles Over Bangladesh*. HarperCollins Publishers India, 2014.

right side of Gun. As Su pushed forward, a bright reflection caught his eye. It was the gleam of metal. He had spotted a Sabre aircraft, descending from 4,000 feet. The Sabre was positioned at 2 o'clock to the formation and closer to Gun's section, approximately 3 kilometres away. It appeared poised for a dive.

'Contact, contact!' Su shouted into his radio.

Young Su sounded excited, yet he remained in control, guiding those in line with the Sabre along the flight path.

'Gun, Don! Aircraft to the right at 4,000.'

Gun couldn't see the enemy aircraft, but Mouse picked up the Sabre on his screen. He climbed and turned right, chasing the Sabre, while reporting the contact to Bagchi. Mouse was ready to engage and fire.

He had closed in, to about 1,200 yards, when the Sabre suddenly veered to the right. While turning, the Sabre dropped speed, allowing Mouse to get closer. The distance between them was 800 yards now. Mouse unleashed his first burst of 30 mm high explosive (HE) rounds with tracers. Su observed the tracers streaking towards the Sabre, but they narrowly missed their target, flying past it.

Massey had missed him. But the chase was on!

Now, a second Sabre sneaked in, attempting to jockey behind Mouse and Su as they pursued the Sabre ahead. In doing so, the second Sabre unintentionally crossed paths with Don and Gun, who were trailing Mouse and Su. The Sabre hadn't seen the two Gnats coming up from behind. Gun spotted the second Sabre on his windscreen as he and Don began their chase. It was going to be a fight to the finish.

As Don chased the second Sabre with Gun, he noticed

a third Sabre trailing behind Gun, flying away from them. Turning right, Don had the tail of the third Sabre in his sights.

The Indians had gained an advantageous position in the duel.

Don worked up a speed of 350 knots and sneaked up close to the third Sabre now. About 150 yards separated the two as they streaked through the air at great speed. Skilled with good reflexes, Don quickly got into a vantage position and fired a burst of twelve rounds. The Sabre wobbled and began emitting smoke. Immediately after, there was an explosion. An elated Don reported the hit to Bagchi on the ground. 'I got him!'[25]

But there was bad news for Don too. As his Gnat closed in on the stuttering Sabre, debris from the explosion shot up and landed on his aircraft, damaging the nose and drop tank, which was leaking fuel.

Gun, who was chasing the second Sabre, called out on the radio: 'Murder, murder, murder!'

He fired a burst, and the enemy aircraft was hit. It caught fire and exploded. Inside his cockpit, Su had seen the canopy flying off through his gun camera. Things happened so quickly that nobody saw the parachute eject.[26]

Two Sabres had been hit. The Gnats were lighting up the sky!

Mouse had missed his target earlier but now he positioned himself behind the enemy plane. Chaudhary, the Sabre pilot, adroitly used a dogfighting move to break Mouse's attack line and forced him to take a high angle-off burst – firing at a target from a steep angle – which missed the target.

Initially outmanoeuvred by a skilled Pakistani aviator, the young Indian pilot quickly regained position and fired a burst at around 700 yards. This time, he hit Chaudhary's aircraft in the port (left) wing. The Gnat was on course to put the finishing touches to the duel, but in an unfortunate twist, Mouse's starboard (right side) cannon stopped firing. Chaudhary's Sabre flew back into Pakistani territory though it continued to billow smoke and caught fire.

Chasing the Sabre, Mouse soon found himself inside East Pakistani airspace. Before bravado got the better of him and realizing the risk, he turned around and flew back into India.

During the duel, Indian radar controllers at Barrackpore were assisting the Indian Gnats by providing them with real-time information and guidance during the aerial engagement. A radar controller plays a vital role in detecting and tracking aircraft, providing critical information for air defence and guiding fighter pilots during engagements. Chaudhary's team in East Pakistan, however, was at a disadvantage because they lacked radar support, but he showed grit and courage by piloting his Sabre back to Tejgaon airfield outside Dacca.

The Wing Commander had escaped despite the damage from Mouse's hit and steered a safe landing in difficult circumstances. He had, however, lost two of his young pilots. To save face, he claimed that he had shot down one of the Gnats. But the truth was otherwise, as no Gnat was lost that day.[27]

It was all over in under three minutes.

The Gnats carried out the victory rollover for jubilant soldiers on the ground. The young pilots – victorious legends even before they had touched down on ground – flew together towards the Dum Dum base. They began their flight as young tyros but returned from the fight as toasts of the nation. Serenading a victory parade in the clear Boyra skies crowned a perfect finale to the battle.

The Boyra Boys were born – that's what the young legends would now be called!

On the Pakistani side, the audacious dives of the Sabres had ended in an unenvious way. They had lost two pilots and the battle. While Chaudhary managed to fly back to Dacca, two pilotless Sabres plunged towards the earth.

The top guns of Boyra

Pilots and POWs: An unusual discussion

A week before the air battle, 4 Sikh had taken control of an area north of Boyra. However, over the past two days (21–22 November) the relentless strafing by the Pakistani aircraft had left them feeling frustrated and powerless.

They watched the dogfight with excitement, hoping that the IAF would deliver a strong blow to the Pakistani aircraft. When the battle ended, and two of the Sabres deployed parachutes, the soldiers were elated. Among the onlookers were Panag and Apte.

One of the parachutes carrying the pilot floated towards Panag's battalion area. Fearing that the enraged soldiers might attack the Pakistani pilot, Panag sprinted to the spot. When he got there, Panag saw that they had knocked the pilot down and were hitting him with their rifle butts. He shouted at the jawans to stop and even shoved a few men aside, throwing himself upon the pilot to save him from being badly injured.[28] Though hurt, the pilot – who was Parvaiz Qureshi – put on a brave face. The other Pakistani pilot, Khalil Ahmed, parachuted to the ground at an adjacent location.

A dazed Qureshi hobbled to his feet. Panag recalls how, despite the shock of being shot down and taken as prisoner of war, he appeared 'stoic and dignified'.[29] The young captain took him to the battalion headquarters and offered him a cup of tea, while a unit doctor attended to his injury.

Panag began the preliminary investigation. His full identification was Flight Lieutenant Parvaiz Mehdi Qureshi, Squadron Commander of 14 Squadron PAF based in Dacca. He had his wife's photo in his pocket. Panag made a list of all his belongings: A watch, a 9 mm pistol, 30 rounds of ammunition and his survival kit. He told Panag that he hadn't seen the Gnats and thought that he was hit by ground fire.

Khalil Ahmed had landed in the area of 1st Battalion, Jammu & Kashmir Rifles, and was apprehended by Lieutenant Jagdeep Sarai.[30] Reconstructing the exact details of the air battle and comparing it with the version released by the Pakistanis, it was inferred that Khalil Ahmed was Ganapathy's victim while Lazarus had brought down Qureshi's Sabre.[31]

Qureshi and Ahmed were the first prisoners of the war of 1971, which was yet to begin.

The two prisoners were blindfolded by Ajit Apte, who used ropes to tie their wrists since handcuffs weren't available. Apte took them to the command post of the artillery battery from where they were to be taken to Calcutta. At the post, their rope cuffs were removed, after which the pilots spoke freely as they sipped tea and smoked cigarettes. Though visibly disheartened by their circumstances, the POWs couldn't help but praise the skilled Gnat pilots.

Ahmed, who was the more talkative among the two, informed Apte that although the pilots had received information about the positions of Indian artillery guns and tank sites, they struggled to find the exact targets based on those maps. Qureshi claimed they were unable to pinpoint the artillery positions even when the anti-aircraft guns were firing at their Sabres.

Apte, on the other hand, claimed that all the diving planes had scared the Indians, making them believe they had been located and targeted. Ahmed, however, explained that diving allowed for a better shot at the gun positions.

How had the Indians managed to camouflage so effectively? Their secret weapon was the villagers. 'We told them to cover the tracks, so they used their ploughs and other tools to make the tracks resemble tilled fields,' Apte explained. Qureshi admitted that, from the cockpit, he had seen hundreds of metres of ploughed fields, unaware they were actually disguised tracks.

The Pakistanis had intended to carry out pinpoint bombing, which involves precise targeting during aerial bombardment. However, due to their inability to locate the target even after diving for a closer look, they repeatedly returned without carrying out the intended attack. Apte was amazed by their audacity.

Night out in the city

While the POWs were taken to Calcutta for interrogations at the command headquarters, the four Indian pilots were soaring high after their victory.

Following the encounter, on landing, they received a hero's welcome in Calcutta. At Dum Dum airfield, a festive atmosphere filled the air. A group of officers and technicians stood eagerly, their attention solely focused on the arrival of the Indian pilots. In fact, at around 1515 hours, the pilot of an Antonov-12 aircraft assigned to the Aviation Research Centre was asked to circle the area when he asked for permission to land at Dum Dum. 'We have four Gnats beating up the airfield and Calcutta at low level for the past ten minutes!'[32]

The roaring engines of the Gnats in the skies brought residents from Park Street to Ultadanga in Calcutta out of their homes and on to the streets. The four ecstatic pilots, fresh from their triumph over the highly regarded Sabres, were showing off their low-level flying skills over a city that would soon welcome them as heroes.[33]

The Gnats were not as advanced as the transonic Sabres, but they embodied the spirit of a classic David versus Goliath battle. Just like the Boyra Boys, the Gnats proved their prowess and cemented their reputation as the Sabre Slayers.[34]

On the brink of war with Pakistan, India had already achieved early success. To recognize their significant contribution, the Defence Minister at the time, Babu Jagjivan Ram, visited Dum Dum along with his wife and Air Officer Commanding-in-Chief Hari Chand Dewan. Sikand, the four pilots, and Flight Controller Bagchi were praised for their commendable efforts.

After the felicitations, Sikand suggested a night on the town to celebrate. That evening, they decided to ditch desolate Dum Dum for the lights of Park Street. Jazz was big in Calcutta in those days and Blue Fox was the place to go to. The well-known Pam Crain was singing that evening when four men in blue overalls walked in.

The manager hurriedly made space for the young pilots at a table. Blue Fox now boasted the four men, known for their heroic battle and victory rolls, that everyone was talking about. Amid the long evening of drinking and dancing, a pretty, young girl came up to Su Soares and asked the handsome young Bandra boy for an autograph. 'I have no pen or paper,' she told him, thrusting her bosom forward and asked him to sign her white T-shirt. Smiling politely, Su asked her to turn around and signed the back of the tee instead.

Several women, armed with pen and paper, took autographs from the young pilots that evening. A decade later, Su found himself attending a course at the staff college in Wellington. At the officers' club, a lady approached him, holding a piece of paper, and said, 'Squadron Leader Soares, this is your signature.'

The Boyra boys: together in Bareilly in 1966

'Yes, ma'am, it is. Where did you get it?' Su asked.

'Blue Fox,' said the lady, now an army wife.

Flying Officer Pradeep Kapoor, along with two officers, was flying from Bakshi ka Talab airbase in Lucknow to Barrackpore Air Force Station to pick up a Dakota (DC-3), a military transport courier aircraft, for war readiness. However, during the flight, he received a message about a change in the flight plan. His new assignment was to transport two individuals to New Delhi.

After the passengers had boarded, and the doors closed, Kapoor asked the pilot who the passengers were. He was told there were two Pakistani prisoners on board. A team consisting of Group Captain Ram, a lieutenant colonel, and a jawan armed with a 7.62 mm rifle had been assigned to accompany the prisoners.

With the aircraft cruising at 8,500 feet, Kapoor decided to pay a visit to the prisoners. A brooding Qureshi, whose head had been bandaged after the pummelling he received from the ground troops, kept to himself. Ahmed was chatty and friendlier, as if he didn't want the past events to dampen his spirits. 'There are lots of discos in New Delhi, yaar. I want to visit them,' he jokingly told Kapoor.

Dinner for the POWs had been sent from Calcutta's prestigious Great Eastern Hotel, and was served onboard the flight. Qureshi declined the food, remaining steadfast even after Kapoor ate a small bite from his plate to assure him it was not poisoned.

Mid-journey, Ahmed asked to use the toilet, which was located at the back of the plane. As he made his way there, Ahmed noticed the armed sentry, assigned to guard the prisoners, had dozed off, with the rifle lying out of his reach. It had been a long day and the cool air on the plane must have lulled him to sleep. Was this Ahmed's opportunity? He could seize the rifle, shoot everyone, tune into the radio compass and fly the aircraft to Lahore.

In the meantime, Kapoor decided to check on Ahmed and began walking towards the back of the plane. Just then, Ahmed appeared.

'What happened?' Ahmed asked.

'I wanted to check what's taking you so long,' Kapoor confessed.

'It could have taken longer . . . see, is that jawan guarding us back there? He has fallen asleep with a weapon alongside him,' Ahmed said, with a smile.[35]

Startled on hearing this, Kapoor woke the sentry and informed the officer accompanying the team. As they took their seats, a thought crossed his mind: if Ahmed had indeed taken the rifle and hijacked the plane, that would have been a turning point for the situation. The mere idea sent shivers down his spine.

A brief and decisive war

By December, the conflict rapidly escalated into a full-scale civil war between the Pakistani military and the Bengali forces supported by India.[36] The PAF continued to violate Indian air-space, even in the west.[37]

On 2 December, at around 1745 hours, Ajit Apte received a parcel from home. The cover of the box featured a photo of Sharmila Tagore, and within was a transistor of the brand she endorsed. There was also a card from his parents for his birthday, which was almost a fortnight away, on 18 December. Apprehensive about communicating if war broke out, Apte's parents decided to send their best wishes ahead of time.

The transistor proved to be a godsend. That evening, Barun Halder and Surojit Sen – whose impeccable diction and deep baritone voices were a familiar presence on All India Radio – announced that the PAF had launched a series of air raids on IAF airbases and radar stations in the western sector, primarily in Punjab, Jammu & Kashmir, Haryana and Rajasthan.[38] Apte's first thought was of his cousin, Flight Lieutenant Pradeep Apte, serving with 220 Squadron and assigned to carry out offensive missions against targets on the western front.

The IAF stood prepared to launch a counterattack in

response to the Pakistani air offensive, known as Operation Chengiz Khan.³⁹

It seemed that the Boyra air battle, which took place on 22 November 1971, had served as a precursor to the war that India entered on 3 December 1971.

Panag and Apte may not be remembered in history as the pilots who shot down the enemy planes at Boyra, but their contribution to the mission's success remains significant. The effective camouflage by Apte's artillery unit held the enemy at bay until reinforcements arrived. Panag's quick thinking ensured that the Pakistani POWs were treated in accordance with the Geneva Conventions.

Unfortunately, not everyone was playing by the rules.

On 4 December, the news came in that Flight Lieutenant Pradeep Apte, flying a HF-24 Marut fighter bomber aircraft, had been shot down over Dhoronāro in the Sindh area of Pakistan.⁴⁰

Pradeep managed to eject safely, similar to Qureshi. However, unlike Qureshi, Pradeep was captured and killed on the ground in Pakistan, in violation of the principles of war.⁴¹

Besides, in a cruel resemblance to the story of P.K. Tayal and Don Lazarus, Pradeep was not slated to fly in that wave of attacks. But the pilot due to fly in the wave fell sick and Pradeep had filled in for him at the last moment.⁴²

Crushed by the news, Apte's mind was filled with questions. What would happen to Pradeep's young wife, whom he had left behind? What would become of the Pakistani pilots captured in Boyra, whom he had recently met? How did Pradeep confront his imminent death in Dhoronāro?

Most importantly, he pondered the nature of war. Was it solely about taking the enemy's life or was there a place for compassion and saving lives as well?

A brief and decisive war between India and Pakistan ensued, altering the geography of the subcontinent. The conflict concluded on 16 December 1971, as the Pakistani military surrendered, leading to the establishment of an independent Bangladesh.[43]

India detained approximately 93,000 civilian and military POWs and accorded soldierly treatment in line with the provisions of the Geneva Convention. On 28 August 1973, India and Pakistan signed an agreement in New Delhi to repatriate these prisoners back to Pakistan.[44]

On the other hand, the fate of fifty-four Indian POWs remains unknown and unaddressed.

Flight Lieutenant Pradeep Apte was posthumously awarded the Vir Chakra for his bravery in the line of duty. His body was never recovered.[45]

And what of the Boyra Boys?

Wing Commander Sikand, serving as the commanding officer of 22 Squadron, was honoured with the Vishisht Seva Medal (VSM) for the squadron's exceptional and commendable service during the conflict.

Roy 'Mouse' Massey, Gana 'Gun' Ganapathy, and Donald 'Don' Lazarus were honoured with the Vir Chakra for their brave actions during the battle. K.B. Bagchi, the battle controller, was awarded the Vayu Sena Medal for his exceptional guidance and direction during the combat.

Though he did not get a gallantry award, Sunith 'Su' Francis Soares played a pivotal role by being the first to pick up the Sabre that day. His quick thinking had a significant impact on the operation's outcome.

1997: Reconnecting with the past

Time had been kind to some of them, thought Don, as he held the letter. Not all.

In the early 1970s, Gun took his own life after suffering a series of personal setbacks.

Mouse went on to become the commanding officer of 224 Squadron (Warlords) and was killed during a routine sortie over Tilpat, Haryana, on 28 November 1983.

Don himself had had a distinguished career[46] and served as the inaugural commanding officer of 102 Trisonics, India's highly classified squadron operating the MiG-25R, an advanced reconnaissance aircraft used for gathering intelligence and conducting surveillance missions. Following his command, he was promoted to the rank of Group Captain and took charge of a major Air Force Station.

Having concluded his air force career and service to the nation, Don was now heeding the call of God.[47]

It was in retirement that Don came across the news that Air Chief Marshal Parvaiz Mehdi Qureshi had been appointed Chief of Air Staff of the PAF in November 1997. Coincidentally, over 25 years ago in November, Don had successfully shot down Qureshi's Sabre in a dogfight.

The prisoner of Boyra had risen to lead his own air force, thought Don.

Don decided to write to his old foe, congratulating him on his appointment and mentioning that Qureshi might not remember their previous encounter *'in the air'*. Don sent the letter, not expecting a reply. However, an acknowledgement did arrive from Qureshi's staff officer thanking Don for the greetings. Don appreciated the gesture and put the letter away, moving on.

That morning, when the postman had handed him a letter, he was indeed surprised. It was a personal missive from Air Chief Marshal Parvaiz Mehdi Qureshi thanking Don for his wishes and complimenting him on the 'fight' in Boyra.[48]

Till date, Don preserves that letter carefully, a timeless reminder of the enduring chivalry among fighter pilots.

Postscript

Mehdi demonstrated not only his courage as a pilot but also his prudence as an astute Air Chief in the years that followed.

In 1998, when Prime Minister Nawaz Sharif decided to appoint Pervez Musharraf as Chief of the Army Staff, Air Chief Marshal Parvaiz Mehdi is known to have advised the PM against the move. Subsequently, Mehdi cautioned the PM against involving the PAF in the Kargil conflict with India. His perspective was that deploying the PAF would likely trigger a retaliatory response from India, escalating the war.[49] Mehdi found himself in opposition to his long-standing coursemate from the 1964 batch, Pervez Musharraf.

As events unfolded, Mehdi's insights proved prescient on both fronts: Musharraf's Kargil venture culminated in defeat, and Nawaz Sharif regretted his decision to appoint him as Army Chief.

*Air Chief Marshal Parvaiz
Mehdi Qureshi*

Part III

Modern Era: Unseen Adversaries, Identity Wars

•────•

Beginning with conflicts in distant lands in the early part of the century, wars crept past the nation's borders over the years, striking our societies and homes. Stories of fighting unseen enemies, slow-drip insurgencies, complex perceptions and deceptive truths entered our living rooms. These stories pose more questions than provide answers.

When families became entwined in a soldier's fight against terrorists during the 2008 Mumbai attacks, a loop that had started thousands of miles away in Europe in WWI had finally come full circle. Stories of wars became a part of us.

•────•

7

A Bloodless Pact to Victory

An Unlikely Alliance

In the months following the end of the Kargil War in July 1999, a few unresolved disputes continued to linger. The incident described in this account occurred in October of that year, a sombre episode in the nation's history.

Pakistani soldiers infiltrating positions on the Indian side had been routed and pushed back, but entire companies from several battalions had made the ultimate sacrifice for the cause. Hundreds of Indian soldiers died fighting in Batalik, Drass and the Mushkoh valley, located in Kargil district of Jammu & Kashmir. The official death toll on the Indian side reached 527.[1]

Gallantry awards were announced for acts of bravery even as flag-draped coffins arrived at airports across India. Families waited to receive medals on behalf of their absent loved ones. The survivors returned, wounded and traumatized.

The LOC stood as a reminder of the uneasy separation between two nations.

Opposing battalions occupied both sides on the formidable mountain range, crouching like brawling fighters gauging their chances of landing a decisive punch, seizing ground and wresting the advantage.

During the ongoing war, radio frequencies were intercepted by both sides, and ground commanders exchanged words, at times, devoid of hostility. Conversations to disagree, clash or de-escalate ambitions weren't uncommon occurrences.

This is an unusual story[2] that unfolded on Marpo La, a crucial ridge overlooking critical supply routes and military positions in the Kargil region.[3]

A conversation about a ceasefire

An exchange was unfolding between the Pakistani and Indian commanders, each referred to by their respective code names: 'Aziz' for the Pakistani commander and 'Victor' for the Indian brigade commander.[4]

'Aziz for Victor . . . please consider the terrain and tough conditions. It will be tough to hold for either one of us.' The Pakistani officer's voice was both shrill and conciliatory.

Victor, the Indian officer, grew impatient, half expecting the Pakistani to initiate peace talks. 'Victor for Aziz – what is your point? Over.'

'Can we agree to halt . . . not attack each other? Over.' Aziz's tense tone revealed the precarious position of his troops on the ground.

Both commanders had a feeling that the conflict was ending, although they couldn't predict when.

'They're in a tight spot . . . those guys. Now they want peace. What other choice do they have?' Victor exclaimed to his colleague, who chuckled in response.

A Bloodless Pact to Victory

On the radio, Victor decided to play along. 'Okay. I agree. Let's not attack each other. Over.'

After a momentary pause, the Indian swiftly laid down the terms. 'You vacate Charlie Two, and I vacate Charlie One ... over.'

Without hesitation, Aziz responded, 'Wilco,' acknowledging the agreement. ('Wilco', a military term, is an abbreviation for 'will comply'.)

Pt 5240, a rocky peak situated on the Marpo La ridge, acted as a barrier between Victor's unit (code-named Charlie One) and Aziz's unit (code-named Charlie Two). Situated on the LOC, this geographical landmark divided the territories of India and Pakistan.

Pt 5240

The tallest feature in the area, and shaped like a knuckle, Pt 5240 stretched over 500 metres at the top. With its steep incline of 80 degrees and a precarious layer of slippery snow, it made for a gravity-defying climb. No one had ever set foot on the feature, which meant there wasn't a foot track available.

Troops on both sides had entrenched themselves for weeks facing each other at Charlie One and Charlie Two, prepared for action, while battles raged in the surrounding areas.

Neither side had moved or withdrawn, as both recognized that their presence was crucial for maintaining control over the Marpo La ridge through observation and firepower.

However, holding their respective positions on opposite sides of the towering feature had become increasingly difficult. Plummeting temperatures, biting wind chill and treacherous rocky terrain compounded the difficulty of attempting an uphill climb to confront a determined enemy.

Both sides acknowledged the formidable difficulties of holding these positions. Wars can unite two fiercely opposing forces against a common adversary – the unforgiving terrain and harsh weather.

That evening, through a radio transmission, two rival commanders – Victor and Aziz – reached an agreement to retreat from their respective positions and relocate to the base of the mountain ridge on their respective sides.

Truce under threat

At the bottom of the mountain ridge, the Indian side featured a levelled valley ground known as the base. The imposing feature prevented either side from having direct visual contact with the other. Defiladed from Pakistani artillery fire, the Indian manpower reserves were stationed at the base along with essential supplies, including ammunition, medical provisions and vehicles. When the action unfolded on the heights, the base became a chaotic scene of medical emergencies and bustling activity. Casualties were brought to the base, while those with more serious injuries were evacuated to hospitals in other locations.

As per the agreement, Major Anand and his rifle company had descended from the heights of Charlie One to regroup at the base. From his position below, Anand lit a cigarette and gazed up at the towering feature that seemed even more imposing from a distance.

He and his men had held Charlie One for twenty days, spending that time tucked inside their sleeping bags and emerging only at night. Surviving the cold, rocky terrain on a diet of biscuits, chocolates and dry missi rotis, they had awaited orders for an attack. Their presence, and position, had already been discovered by an enemy sentry who had noticed the glint of a light machine gun barrel in the sun. Anand anticipated orders for the capture of Pt 5240 to arrive soon.

'Taking the peak will mark the end of the war,' was Anand's favourite line to keep himself motivated during the twenty consecutive days of battling the weather in the heights.

When orders finally did arrive, the men were dismayed to hear that they were being told to return to the base. The descent proved to be more challenging than the trek up, as the men had to navigate the steep and narrow slopes while carrying weapons, ammunition and heavy supplies on their backs. This gruelling journey left them with sore bodies and trembling knees.

Mornings at the base were typically uneventful. Men lined up for tea, breakfast and the latrines in such a predictable order that it was jokingly referred to as the 'morning drill'. 'Ram, Ram, Sahib!' Rifleman Param Singh said while handing Anand his morning chai.

Tea accompanied by a Wills Navy Cut smoke first thing in the morning was a ritual for the Major.

Despite the night's rest, his back ached from the previous day's trek down. Embarking on the next phase of his morning drill, he headed towards the latrines.

Anand had spent two years in the mountains with this unit and shared a strong bond with his men. In their company, he ate langar dal, vegetables and junglefowl, slept under the open sky in a sleeping bag, listened to their stories of home, and participated in their pranks and silly jokes. In the officers' mess, Anand often quipped that the morning drill served as his only personal retreat within the company.

That morning, his private place was usurped.

'Ram, Ram, Sahib!' Naik Gulab Kumar, his radio set operator, responsible for maintaining communication equipment and relaying messages, waited impatiently outside the makeshift toilet. A minute later, a handset was thrust in through the tent opening.

'Oye, wait,' Anand hollered.

'Big Tiger calling, Sahib!' Gulab sounded nervous. (In the army's radio telephony procedure, commanding officers of a battalion are referred to as 'Tiger'. A brigade commander or commander of a higher formation is colloquially called a 'Big Tiger'.)

'Good morning, Anand. Listen to me carefully . . .' Victor spoke with an unmistakable tone of an impending disagreeable order. He pointed out the tactical advantage they had over the enemy, as all three positions – Charlie One, Charlie Two and Pt 5240 – were unoccupied. 'The time to strike is now!' Victor said gleefully, tasking Anand's men with climbing Pt 5240 to surprise and outwit the retreating enemy. 'Now?' Anand asked, referring to the agreed terms on both sides.

'Screw the agreement with the enemy, Anand,' Victor said, matter-of-factly. He added, 'Son, in wars, we pick opportunity over trust.'

Momentarily disconcerted by Victor's childlike enthusiasm, Anand hesitantly responded, 'Yes, sir.'

'I would like to see you on top by 1830 hours today,' Victor said, ending the call. The matter had been settled.

The news set off a pandemonium at the base.

Men scampered around, gathering weapons, ammunition and stores for the operation. Anand heard the discontented whispers and smiled inwardly. He realized he was not alone in feeling bothered by this turn of events.

'Roti kha ley, sab theek ho jayega (Have some food, everything will be fine),' the Company Havildar Major (CHM) hollered routinely, though his voice was filled with an uncommon display of empathy.

In the army, no one marches on an empty stomach, and food serves as a suitable remedy for complaints.

The CHM read out the instructions and shortly after, thirty men outfitted with haversacks and packed breakfasts, and armed with weapons and ammunition, began their trek up the mountain. Dusk had settled, and the silhouette of the rocky peak speared into the greying sky. Single file, the troops climbed the same route they had descended a day earlier. 'War is about going up and down the hills. When you get better at it, you win,' Anand said to Subedar Mir Ali, junior commanding officer (JCO), as they walked.

In due time, all three targets came into view: the smaller jagged features of Charlie One and Charlie Two on either side of Pt 5240. The men had reached the final phase of their climb when the sun set behind the mountains. 'Last 200 metres . . . your movements will be tactical,' Anand reminded.

On his signal, they changed their pace to well-practised stealthy movements, bending, slithering, sneaking forward, using rocks to conceal, determined to keep the news of their climb to themselves.

Close to the summit, Anand raised his fist and signalled for the men to stop. The fading light made it difficult to see, but he managed to observe Charlie One and Two through his binoculars. They seemed empty and devoid of activity to Anand, leaving him to wonder if the enemy had made a similar move, violating their agreement.

Just then, a hail of gunfire from a universal machine gun ripped through the darkness and bullets ricocheted off the rocks, dangerously close to his men. Just as Anand had suspected, the Pakistanis had been lying in wait.

Anand's men scampered for cover, fortunate not to be taken out at the start. Sparks from the next round of fire helped Subedar Mir Ali, crouched behind a rock, identify the enemy position and their disadvantage on the ground. 'Thodey peechey aur neechey hain hum se (They are positioned slightly below us and at a distance).'

The enemy was positioned at a greater distance from the summit than Anand's men, and the latter were determined to keep it that way.

Ali, a former weapons instructor at the Infantry School in Mhow, Madhya Pradesh, knew exactly which weapons to use in difficult situations. He gestured for Havildar Babu Banik to position the automatic grenade launcher (AGL) at a designated spot, while Havildar Ram Singh set up his MMG. The men affectionately called Banik 'Mendhak (frog) guruji' due to the frog-like base of the grenade launcher. Banik placed the base with a decisive thud, resembling a cricket fast bowler preparing for his run-up. Gunfire erupted from the flank, producing a loud burst of sound. Ram Singh's MMG opened up, and it became a 'jugalbandi' (duet) as Mendhak's grenades complemented the whizzing bullets of the gun.

As Ram Singh and another gunner fired to keep the enemy under control, Anand and Ali quietly led their men towards the summit.

Anand acknowledged the wisdom of their commander's words about making smart tactical choices in war. 'Son-of-a-bitch, the old man was right,' he thought. Any resentment he had for the commander dissolved in that moment.

'Vijay...Vijay...Vijay!'

The climb to the top demanded mental resilience and robust physical fitness. The gritty group had tackled the steep stretch that led to the top, with each soldier carrying over 20 kilograms of weapons, ammunition and equipment. A few of them had trained at the Indian Army's High Altitude Warfare School (HAWS) in Gulmarg but negotiating a steep climb with poor visibility, slippery bends, rarefied air and a waiting enemy was a tough challenge for the group before the battle had even begun.

On reaching the summit, Anand checked his watch. 1955 hours, it was good timing. He and Ali exchanged a high five. 'Sahib, well done!' Ali said.

It was back to business after that. Anand radioed the base to report their progress.

'Anand for Victor ... Anand for Victor ... Over.'

'Go on, what news? Over,' Victor's voice had lost its boldness.

'Anand for Victor. Vijay ... Vijay ... Vijay ... Over,' Anand exclaimed, using 'Vijay' to denote victory. His excitement was hard to hide, and throughout the rest of the conversation, both parties swung between caution and joy.

Victor asked, 'Where are they? At Charlie Two?'

'They are around but pinned down,' Anand said confidently.

'Keep a watch. They might attack,' came the measured response.

The soldiers spent a restless first night at the summit, awakened either by gusts of wind or the sound of enemy movements. A cold, bruising breeze stabbed at their noses and cheeks. High up in the mountains, dawn arrived early. Anand, positioned at an elevated point, now had a clearer view of the entire ground, including the direction from which they had been fired upon.

Now in a defensive position at the top, he considered the possible routes the enemy could use to launch a counterattack and drive them out. His thoughts went back to his time in the Young Officers Course at the Infantry School in Mhow, where his ideas on adopting 'all-round defence' were criticized by his instructor.

'Thank God I was a little crazy!' he mulled.

Although chuffed that the men scaled the unclimbed peak quickly and kept the enemy down, he was aware that a lot was left to be done. To start with, he and the men had to find positions to conceal themselves.

Once the team was deployed in position, he sited the automatic weapons to cover the likely routes of an enemy attack.

'Ram Singh aur Mendhak sahi jagah hai, sir. Aur raastey cover kar liye, sir (Ram Singh and Mendhak are in the right position, and we have covered all other routes too),' Ali assured.

For Anand, all-round defence meant having the right men in the right positions with their weapons sited. All-round confidence came from trust in the men who manned his special weapons, in this instance, Ram Singh or Babu 'the Mendhak' Banik. Mir Ali's support served as validation.

The day turned out to be an anticlimax. It was mostly quiet, a stalemate where neither side fired but instead sized each other up. The stalemate was finally broken, but it wasn't by fire. Instead, it was an unexpected move no one had anticipated.

One of Anand's scouts, tasked with keeping watch on the route towards Charlie Two, gave a shout. 'Charlie Two ke peechey (Behind Charlie Two)!'

Two figures emerged from behind the rocks below Charlie Two.

Ali had kept an eagle eye on the area, in a bid to locate hiding spots. His eyes lit up. 'Pata tha ... (I knew it),' he said. Ram Singh took aim at the machine gun.

Anand took the binoculars. 'Two men are holding up a white flag,' he said.

'White flag, sir?' Ali was puzzled.

The two Pakistani soldiers had tied a snow jacket to a pole, which they now waved at Anand's team.

On the Indian side, there were differing opinions on the meaning of this gesture. Ali was sure it was a trap, while Ram Singh was ready to fire.

Anand had kept his gaze fixed on the two men, observing their movements closely through his binoculars. He raised his right palm and asked his men to pause. 'Nobody moves. Keep looking. Be ready,' he said tersely.

As the two figures advanced, the white flag appeared even higher than before. Anand's team was perplexed at the turn of events. Carefully, the enemy soldiers moved closer, seeking safety on the narrow ledge even as they faced the pointed weapons. They had walked the longest 100 metres of their lives. Clicking sounds of the cocking of rifles greeted their advance, and the situation became unbearably tense.

'No one fires,' Anand shouted, as the enemy soldiers reached the halfway mark to them.

Anand stepped aside to confer with Gulab, the radio operator, and Ali. While Anand trusted Ali's instincts, which told him it was a trap, he remained unsure. 'The subedar is usually right about these things,' Gulab chimed in.

'That's why you will come along with the radio,' said Anand, with a smile that betrayed his nervousness.

A surprising encounter with the enemy

Neither Anand nor Gulab had carried weapons. However, Anand had given Ali and Ram Singh clear instructions on how to proceed: *Keep the men in your sights and finish them off if they pull any stunts.*

The Indians followed a rocky spur descending from the higher point of Pt 5240 towards the two Pakistani soldiers, who appeared a few minutes later. The flapping of the snow jacket, carried by the breeze, made the two soldiers resemble flagbearers in a ceremonial parade.

One of the soldiers looked around 30 years old and had a beard, while the other was thin and had a pale complexion.

The latter, who had the look of an officer, stepped forward to speak.

'Hello, who is the commander?' he said, straight to the point. Gesturing towards the position below Charlie Two, he added, 'I am Captain Shoaib, commander of the party down there.'

Anand realized that their position was further down from the place they had assumed. Knowing that his machine gunner was ready in the rear, he quickly scanned the two officers for hidden weapons. By then, the tension was thick enough to cut with a knife. Anand and Gulab exchanged a brief glance. The Pakistanis looked uneasy.

Indian officers are prohibited from disclosing their true identities to the enemy, so Anand gave them a false name. 'Major Sachin, the commander.' Anand assumed a cheeky alibi laced with a cricketing comeback for Shoaib.

'Oh, hello sir, how are you?' Shoaib smiled and switched to

a more cordial tone commonly used when addressing seniors in rank and service.

'You guys keeping us busy, Shoaib.' Anand extended a friendly overture as he opened up the conversation.

The young Pakistani officer got straight to the point. 'Sir, you have been occupying our area,' Shoaib said matter-of-factly, holding a map and pointing to the LOC running through the Marpo La ridge. 'Here . . .' he stated, a sense of satisfaction evident on his face.

'Listen, Shoaib,' Anand responded firmly, not willing to be drawn into a lengthy conversation. 'That is your perception of the LOC. But if you look at the maps signed by the Director-General of Military Operations of both India and Pakistan in 1972, Pt 5240 is clearly within our territory. Just so you know, we are standing in Indian territory, and you are here without a visa. So, I suggest you move your group back.'

Shoaib grinned at the visa comment and then shook his head in disagreement.

'What, you don't agree? What about the hundreds of lives lost when you came and illegally sat inside our territory?' a riled Anand countered, the wounds of war still fresh in his mind.

'Sachin sir, I will convey your position to my seniors. They asked me to share our position with you,' said a disappointed Shoaib with a shrug.

'Was it their view or yours?' Anand couldn't mask his irritation.

'Our view . . . Pakistani view,' Shoaib stubbornly maintained.

'So, tell your Aziz there is a map Pakistan signed in 1972.'[5]

After that questionable exchange, they returned to their positions. The meeting had left a touch of bitterness.

Two officers share a smoke

The next morning, Anand asked his sentries to look for signs of movement on the enemy side. 'They may try some tricks... be ready,' he warned. The entire group was on high alert, with the machine guns and grenade launcher ready to pounce on the slightest provocation.

Close to noon, Ali spotted a tall figure through his binoculars. He was carrying a white flag, but not waving it as wildly as before.

'Just one with the flag, sir.'

'One? What about the other guy? Doosra (second)?'

'Nahi sir, akela (no, he's alone). Only one... officer. That's why he is waving the flag gently.' Ali chuckled as the enemy soldier drew closer.

Despite everyone's apprehensions, Anand chose to meet Shoaib alone. 'Surely, a trap... pukka (definitely),' said Gulab, who insisted on accompanying him.

Anand put a hand on his shoulder. 'Gulab, follow me... but stay far behind.'

Taking the foot track along the spur, Anand kept his eyes fixed on the figure ahead. He could sense the presence of other watchful eyes observing them from a distance.

Shoaib greeted Anand. 'Morning, sir.'

'It's afternoon... or did you just wake up? And do you have a reply to my question?' Anand asked.

'No and no. I haven't heard from my boss.' Shoaib chuckled loudly, in a more light-hearted mood compared to the previous day.

'Even Victor hasn't heard anything from Aziz for a while now. Why did you want to meet?' Anand didn't mince his words.

'The sun is out, so I thought of taking a walk,' said Shoaib. Suddenly, a hint of tension appeared in his demeanour, and Anand couldn't help but wonder if he had missed something. A sense of unease enveloped him, and he felt relieved knowing that Gulab was nearby for backup but at a distance where he couldn't eavesdrop.

'Actually, sir, Aziz's reply is likely to be negative. So, I thought of coming here and meeting you . . .'

'That's interesting. We might open up with our weapons. Did you ever think about that?' Anand said curtly.

Shoaib gestured towards the white flag as indicative of why there wouldn't be any firing.

'Sir, do you care to smoke?' he asked.

Anand was curious now. 'Why smoke here?'

'Maybe there, sir. Perhaps more comfortable . . .?' Shoaib wanted to break the ice and pointed towards a pair of large rocks nearby.

Gulab had been eyeing them from a distance. He had overheard parts of the conversation but couldn't follow the English language close enough to learn anything.

He thought of the village elders, the wise old men, who had said that a disagreement was often less intense when people sat down to discuss it rather than standing up. Now, Gulab saw it happening right in front of him.

The Pakistani officer sat down and lit his cigarette.

From Gulab's position, the conversation was barely audible now. 'Looks peaceful. *Shanti hai*,' he thought, without taking his eyes off them.

'So, how long have you been in the army, sir?' Shoaib took a deep drag of his cigarette.

'Twelve years,' said Anand. The number was actually ten, but his instincts told him to lie to the enemy. 'And you?'

'Two years, sir.'

'You look young.' Anand thought, he isn't lying.

'How is the war going for you, sir?'

'You started it. Not you . . . your bosses. And see the mess. So many killed . . . all your guys who infiltrated have been killed. How was the defeat?'

'I lost some good friends, sir. I haven't spent much time in the army and don't understand strategy. Lost good friends . . . that's all I care about.'

'You have been defeated. And you've lost friends.' Anand's tone was disparaging.

'War happens . . . for so many reasons. We couldn't have stayed in these areas forever, sir. You too lost many men trying to push us back,' Shoaib said defensively.

'So . . . what did you achieve?' Anand asked, trying to understand what the young Pakistani army officer had on his mind.

'We? Achieve? Not sure . . . We as in you and us?' Shoaib smiled wryly as he looked at Anand, then gazed out at the mountains. 'So many losses. All for these barren mountains, sir?'

'Seen barren mountains before?' Anand tried to make the conversation more casual.

'No, I grew up in Karachi and Dubai earlier for a bit. City slicker . . . And you, sir?'

'Chhindwara . . . you know where it is? I bet you don't,' said Anand, referring to a town in Madhya Pradesh. He had concealed the truth about his home town, which was Patna.

Shoaib shrugged. 'West India, sir?'

'Now, now, you're firing in the dark! Maro goli (shoot)!' Anand grinned, and then they both broke into laughter. 'Look up the map . . . look it up,' he continued in his best senior officer voice. 'No, in fact, don't. There's a good chance you may not find it!'

They went back to their positions, wearing playful grins on their faces.

Gulab was waiting for his boss. 'Sir, Chhindwara?' is the first thing he asked Anand as it's the only word he could decipher from their conversation.

'Yes, Gulab.'

'Chhindwara? Why did he ask about me, sir?' Chhindwara happened to be Gulab's home town, which Anand knew of.

'He didn't. Not about you. About the place.' Anand smiled, leaving Gulab dumbfounded. Chhindwara had never figured in any of Gulab's discussions earlier with his mates. And now, of all places, it came up in an India–Pakistan chat.

Navigating difficult decisions

A week passed by. A strong chilly breeze blowing on the heights made it difficult to cope with the cold evenings.

On the other hand, an almost warm camaraderie had developed between the two officers who had been meeting often.

Their routine went thus: the white flag went up on Shoaib's side. Anand set out to meet him. The rocks became a regular venue.

'Any news from Aziz?' Anand inevitably asked each time.

A Bloodless Pact to Victory

One day, Shoaib pre-empted and said with a chuckle, 'Sir, aaj Aziz ne kuch nahin bola (Aziz said nothing today) . . . I replied before you asked. This question is your opening delivery of each match. Outswinger!'

'You are the bowler, Shoaib.' Anand doesn't waste a comeback.

They exchanged playful jokes, discussed topics ranging from cricket to Bollywood heroines like Kajol and Madhuri Dixit, and shared stories about the food in their respective cities.

The 1999 ICC Cricket World Cup coincided with the war, and despite India's early exit from the tournament, they had emerged victorious in their match against Pakistan.

'Sir, we did better than you, you must agree . . . and we are a better team,' said Shoaib in jest. 'But you guys have lost to us again in the World Cup. One on one, we always beat you . . . 1991, 1996 and now,' Anand countered playfully, speaking for his alias, the famous Sachin Tendulkar.

'We acknowledge your best batsman, Sachin sir!' Shoaib laughed.

One day, Anand brought a quarter of Old Monk rum in his hip flask. They took a swig each. By now, Anand was comfortable enough to occasionally cadge a cigarette from the Pakistani officer. Shoaib had brought chocolates. 'These are American,' he informed.

They kept the war out of their conversations. They agreed about the weather and survival; there wasn't much to argue about and the bonhomie was just fine.

'Sir, how long do you plan to be here?' Shoaib asked.

His query, which alluded to the war and their current status, came at a time when strong blizzards had begun blowing over the cliffs, making their positions tougher to hold.

'Can't say, can't tell. Not supposed to.' Anand had put on his military hat.

On that day, Shoaib appeared to have reached the end of his tether. 'Do you think it's possible for both sides to continue like this indefinitely? Just sit and survive? I can speak for myself and my men. Sir, we are sitting on a rock with nothing around us. There's just a stove; no place to eat, defecate, wash . . .' Shoaib said, his frustration evident as he let out a curse.

'So, what does Aziz say?' Anand asked. He hadn't mentioned Aziz in a while.

Shoaib ignored his question, and instead asked, 'Are you married, sir . . . ?'

'No, why?'

'You must be older. Did you never think of marrying?' Shoaib pressed on.

'I haven't found the right girl.' It was not exactly a lie. Before he was posted to Kargil, Anand had ended a relationship. 'You ask too many questions, young man. Tell me about yourself. Your turn.'

'Sir, I have been engaged for almost two years now. She has been waiting for me. I have known her a long time,' Shoaib revealed.

'And yet, you're mucking around in such a godforsaken place. Why the delay, shehzada (prince)?'

'I promised her that we would marry once I joined the army. And then the war happened. Now, I've sworn to marry her once the war is over,' Shoaib said gravely.

'So, you miss her ... huh?'

Shoaib nodded softly.

'Toh jaao (Then go). Go get married. The war is over. Chal (come), light a smoke and take a swig of Old Monk.' Anand patted Shoaib on the back as he handed him his hip flask.

'No, I can't just go away and marry, sir. Aziz is going to make me wait forever. I am fighting his battle, you see. Just look at this place where my men are sitting,' Shoaib vented. He took a swig from the flask. 'I don't want to let my men down or lose them.'

'Your men, your women – lots to think about, yeah? So, what do we do about it?' Anand lit another cigarette which they shared. 'You better leave this place before the cigarettes and rum are over, Devdas,' Anand joked, making a reference to the popular Indian lovelorn character from fiction.[6]

'Look, what is Aziz's plan? You are here to attack us there,' Anand said, pointing towards Pt 5240. 'Will you be able to accomplish it with the manpower at hand? You're smart, you know how things stand. We are sitting on top and tactically in a better position to outmatch you.'

Shoaib was listening intently to the Indian company commander. Anand continued, 'Many people will die. Your people ... there will be unnecessary losses on both sides. Can you tell Aziz he is setting you up for the loss of your men?'

'Sachin sir, will you be able to ask your boss a similar question?' Shoaib passed the cigarette back to him.

Anand gave him a weak smile and thought that the young man knew how to hold his ground. Shoaib had thrown down the gauntlet of moral courage. Anand took a long drag on his Wills Navy Cut but remained silent, choosing not to respond.

Shoaib tried another tack. 'Look, sir, Aziz is on the back

foot. His plan hasn't worked, so there is a possibility that I may be ordered to attack the top. But with the ceasefire in place, I don't think there are too many options. So, we will wait it out till we get a chance.' He is sure his men will die. 'Sir, why more deaths here ... and for no reason?'

His candour surprised Anand, who reiterated his point. 'Our counterattack can get to Pt 5240 faster than your reinforcements and our artillery will prevent you from holding your position there.' It was well established that the topography and ground layout favoured India's rear troops, compared to Pakistani reinforcements.

'If Aziz wants you to stay here and build defences, I reckon you already have the right reply for him. Consider your current position, Shoaib. How long can you remain there? Build defences on that ledge? Who will let you do that? We can clearly see your path from your base to this hill. The moment you start bringing up equipment and construction materials, we will bomb you to your graves.' Anand's tone carried empathy, but he remained resolute in his stance.

A long silence followed as they took a swig each of the Old Monk. Anand slid an arm around Shoaib's shoulder, and calmly said, 'Look, you are the commander on ground. You have done a brave job keeping us on our toes, both now and earlier. You have done your nation proud. There is no one better placed than you to take a hard call.'

Shoaib appeared lost in thought, contemplating whether his enemy was offering him better advice than his own superiors. Was shahadat (martyrdom) more worthwhile than saving the lives of his comrades from futile deaths?

'You are the captain of your country, Rawalpindi Express.' Anand grinned, alluding to the nickname of Shoaib Akhtar, the Pakistani pace bowler. The playful reference lightened the mood, and both officers shared a laugh.

'Give me a day to come back on this, sir.' On that serious note, Shoaib departed while Anand stood quietly and watched him go.

The next day, the meeting request arrived earlier than usual, before dawn.

After reassuring Ali, and declining Gulab's suggestion to accompany him, Anand set out to meet the enemy.

With hands on hips and legs firmly planted apart, Shoaib displayed a new-found sense of decisiveness. After shaking hands, the two men delved straight into the matter at hand.

Shoaib said, 'Sir, we can't hold on to the position and my team believes it's not their battle any more. This is Aziz's war and driven by his ego . . . and he doesn't seem to have a plan.'

Anand was direct. 'Should we render your position untenable, then?'

'Yes sir,' Shoaib replied, and as Anand looked into his eyes, he saw courage, patriotism and something even more extraordinary – a humanity that the ravages of war hadn't stripped away from him.

Confronting foes while supporting a comrade

Back at his location, Anand grabbed the radio handset. 'Anand for Victor . . . Over.'

'Victor for Anand . . . how are we?' a voice responded.

'Anand for Victor . . . enemy seen . . . Golf Romeo 864486 . . .

Over,' Anand replied, using military alphabet to communicate the enemy location with reference to the map.

He received an instant response. 'Victor for Anand . . . engage the enemy,' said the eager voice.

'Wilco.'

Anand appeared calm on the outside, but inside, his emotions were in turmoil. He walked up to Naik Ram Singh's MMG and stood behind it. 'Ram Singh, can you see the rock, 3 o'clock to the enemy position?' Anand pointed towards a 'rocky bump' around 300 metres right of the enemy position. He stood up and smiled. 'Let's shake them up.'

Ram Singh was nonplussed. However, he had received an order, so he adjusted his weapon, aligning it with the target. The range was set at 250 metres, and the cocking handle pulled back, ready to fire.

The men threw themselves on the ground and scampered to find positions for the firefight that would follow.

'Enemy movement sighted. Take position,' Anand commanded.

The ammunition belt rolled into the firing chamber of the MMG and a hail of bullets flew towards the target. The grating gunfire of the 7.62 mm machine gun reverberated in the gorges, a booming echo that carried the sound farther, funnelled by the tall, rocky walls of the massifs.

Gulab peered through his binoculars and immediately reprimanded Ram Singh. 'Ram Singh!! Kahaan maara (Where did you aim?)!'

'Chupp . . . (Quiet)!' Anand shot back.

Ali leaned towards Anand. 'Sir, it's a little wide. Gulab is right.'

Anand ordered, 'Durust fire . . . repeat.'

Ram Singh's finger lingered a little longer on the trigger. The echoing mountains roared back in resounding approval as he fired again. The booming cacophony grew louder with each sustained burst, drowning out Anand's throat-bursting approvals.

The echoes of the one-sided gunfire reached nearby friendly positions in the rear. A message crackled on the radio set beside Gulab. Victor was on the line. 'Big Tiger,' Gulab shouted out to Anand.

Victor informed Anand that they heard intense gunfire and were aware that fighting was taking place in the nearby area.

'Fire at them ... use the mortar ...' Victor ordered.

Anand responded, 'Negative, sir. No mortar. Over.'

'Are they advancing towards your position?'

'We are engaging the enemy ... they are firing too ... Over,' Anand said. Briefly recalling Shoaib's plan to attack the crest, Anand swiftly dismissed the thought.

'Take them out, son ... Good luck.'

'Wilco. Over.'

Ram Singh, Gulab and Ali sensed that something was not right. Anand, driven by his own agenda, had meticulously planned the mission.

He ran towards Mendhak. 'Prepare the AGL and wait for my order, Banik.'

Mendhak pulled back the handle of the grenade launcher, his strong arms made it look easy. 'AGL ready, sir.'

'See the nullah (waterway) ahead?' Anand sounded grave; he couldn't afford to get this wrong. He had noted Shoaib's position and believed the nullah was about 500 metres beyond it.

'Seen, sir,' said Mendhak, thumbs on the AGL's firing lever.

'Three rounds . . . fire!' Anand's command was louder now.

Mendhak aligned the sight, adjusted the elevation and raised the barrel for greater distance. Three rounds were fired. A muffled thump echoed in the far distance. Smoke billowed out of the nullah as the shells exploded after their high trajectory flight. The AGL rounds fired at the nullah arced over the enemy's position and plunged into the ravine behind them, detonating upon impact.

The anticipated command was issued. 'Durust (second round) fire.'

By this point, Victor was requesting minute-by-minute reports about the situation over the radio. Unfortunately for him, Anand was now fighting his own war, and not Victor's.

'Can you see them?'

'They have taken cover, sir.'

Radio telephony procedures were abandoned.

'Problem is, I can't see them . . . there is too much dead ground . . . Over.' Anand paused, then said, 'Intermittent firing from their position . . . Over.'

He received his orders. 'Don't let them advance.'

Four more rounds of the grenade launcher were lobbed over the enemy position into the nullah. Thick clouds of smoke built up, creating a dense fog in the far distance. The MMG opened up alongside the grenade launcher. It became a full-scale display of firepower under Anand's command.

The mountains roared and roared, evoking alternating feelings of excitement and panic in Victor. There was also a niggling worry regarding Anand's position.

But Anand was steering the situation now, and he knew exactly what would follow next.

The Big Boy enters the fight

In the distant foothills of the mountains, an artillery regiment was preparing to provide operational support. Gun crews deftly handled the levers of massive guns, loading shells into the barrels, with the automatic loading mechanism taking care of the rest. 'Fire!' bellowed Major Bikramjit Rana of the artillery unit.[7]

The battle had escalated now.

Victor was clear that the enemy wouldn't be allowed to occupy Pt 5240 and was ready to use the big boy: the Bofors gun.

The 155 mm FH 77-B Bofors gun, a controversial purchase from Sweden in the 1980s, had proved to be a mainstay in the Kargil War. Indian infantry officers directed its devastating strikes on Pakistani positions, breaking their resistance throughout the conflict.[8]

The Bofors battery crew plugged their ears. A long whistle pierced the sky, announcing the release of the ultimate weapon to quell the chaos in the mountains. Shells rained down in the distance, erupting in a thunderous display. After a furious spectacle that lasted mere minutes, silence returned to the scene.

Through his binoculars, Anand saw the familiar figure of Shoaib with his white flag. He waved his hands, and tried to say something, but was too far away to be heard. Anand saw him point towards the nullah and understood his concern.

Calling in fire support from the artillery guns placed behind, he said in hushed tones, 'Anand for Bikram . . . add 150 . . . one round gunfire.'

The calculation took a few seconds. 'Fire!' Bikram's voice boomed through the megaphone.

The shells exploded deep into the nullah. 'Good shot,' Anand shouted. He kept his communication precise. 'Repeat ... over.'

Over the next half an hour, successive salvos soared across the skies as the Indian troops unleashed a massive – and intimidating – exhibition of firepower for the enemy. 'Target neutralized ... stand easy ... Over,' Anand signalled. The soldiers surrounding the artillery let out a triumphant war cry, pumping their fists in the air.

The battle had been won.

It was time for Anand to inform the boss. 'Anand for Victor ... target neutralized ... enemy is moving back,' he said.

'Enemy moving back?' Victor sounded curious.

'Yes ... Over.'

In the meanwhile, the troops had taken cover in anticipation of an enemy retaliation.

Anand waited a little longer than the rest. After around twenty minutes, a familiar figure appeared in the distance.

In trackpants and a T-shirt, Shoaib had hurried to their meeting point. He hugged Anand, and said, 'Sir ... sir ... thanks, it's worked.'

'What happened to your artillery? And Aziz?' Anand wanted to know.

'You guys bombed us badly ... I radioed him about the constant heavy fire we were under, and that our small team was in no position to capture the top.'

'And then ...? What about your artillery?'

'I had told a senior staff officer that, in the event of retaliatory fire, it would be difficult for my troops to withdraw or advance. I think I convinced him. He and the others got Aziz to agree. I-I-If this patrol had been w-wiped out d-due to foolish c-circumstances, th-there would have been no o-o-one capable of a-a-answering for it.' Shoaib, out of breath, had begun to stammer.

'Calm down. You have saved 30 lives.'

'Twenty, sir. 20.'

'What! You expected to beat us with 20! Shoaib, man, you're sharper than your senior lot down there. I could put the pips on you right now,' Anand retorted, using a popular military phrase to imply Shoaib deserved a promotion – new pips or insignia – for his actions.

They both laughed.

'But Sachin sir, I think I could've taken a better shot. Your artillery shell landed too close. Sir, that could have killed us. We are better at arty than you guys. Any day,' Shoaib teased.

'Now, now, your piping ceremony is cancelled. But you're granted leave. Go home, sonny boy. Take your big wicket, Rawalpindi Express,' Anand said, as they shook hands for the last time.

Shoaib appeared calmer now. 'Sir, I am going on leave . . . to get married.'

'Congratulations, young man. You will finally hold your promise.' Anand took out the hip flask and shook it a little. 'Should be enough for us both.' He added, 'In war, we make our choices. But on a rare occasion, very rare, the enemy chooses one for us.'

They raised a toast to one another. 'Inshallah – till we meet in better times.' Anand held out the flask for Shoaib, who took a swig from it. Anand then drained the flask.

From his pocket, Shoaib took out an envelope and handed it to Anand. 'Open this when you leave here.' He also thrust a packet with a few cigarettes at him. 'There are a few left in there, for you to remember our smokes together.'

They headed back to their positions. As he turned the corner on the spur, Anand caught Shoaib waving to him. He raised his hand in acknowledgement.

Back at the camp, Anand pulled the envelope out and opened it. On a piece of paper, Shoaib had scribbled: *You are invited for the wedding of Captain Shoaib Hassan and Ms Shagufta Khan on 10 November 1999. Look forward to your presence and blessings.* Below that, he had written: *Sachin sir – Shagufta and I thank you for everything. You gave us our lives together. May Allah bless you.*

Anand lit one of the cigarettes and shut his eyes. It was all very quiet. Nothing had changed in the cold mountains around him.

Just then, the radio crackled with Victor's excited voice. He had called to congratulate them.

Postscript

The Kargil War was a result of General Pervez Musharraf's plan to occupy strategic heights in Indian territory between 11,000 and 17,500 feet.

Pakistani army intruders had illegally occupied several peaks in the sector before being pushed back. A report claimed that Pakistani troops, upon detecting Indian occupation at 5240, occupied the neighbouring peak Pt 5353 by November 1999. Pt 5240 (which is in Indian territory) is now occupied by the Indian Army. Pt 5353 is said to be occupied by Pakistan.

After the war was over, Aziz, the Pakistani commander whose plan of sabotage was upstaged, faced a court of enquiry and disciplinary action for his decisions that led to Pakistan losing Pt 5240 to India.

Shoaib Hassan went on to settle overseas after a distinguished career with the Pakistan army.

8

Warrior's Code of Courage

Leading with Heart and Honour

In June 1997, a bus made its way into a village in Bongaigaon district, situated in the north-eastern state of Assam. On that muggy afternoon, the air hung heavy, the silence only broken by the incessant barks of weary pariah dogs, too exhausted to even bother chasing after the passing vehicle.

A local man on the bus signalled to the driver, bringing the bus to a lumbering halt. A trusted intelligence source, he had solid information about militants hiding in a village and had directed the bus, with Indian troops aboard, to it.

A close collaboration between informants and the military existed in the 1990s, the peak of Assam's brutal insurgency. The United Liberation Front of Asom (ULFA), grown from a rebel movement in 1979 to a militant separatist organization, had accelerated its demand for the sovereignty of Assam. By 1990, ULFA had been banned by the Government of India.[1] Afterwards, separatist rebels unleashed a wave of terror, carrying out killings and kidnappings that specifically targeted Assam's tea industry, officials, important installations and business activities. In response, the Indian Army launched Operation Rhino,[2] an assault on the heavily wooded forest where the ULFA headquarters were located.

Officers had been actively seeking and searching for militants, diligently scouring hideouts, when they received a crucial tip-off about a house near Bongaigaon.

Militants encounter a formidable team

On the bus was Major Deependra Singh Sengar, his team of commandos from the 21 Para Special Forces (SF) and a few policemen. The 21 Para SF had gained expertise in operating in the areas under the Eastern Command[3] and kept a watchful eye on the village through the bus windows.

Having just returned from his annual leave to his posting in Bongaigaon village, Sengar had been approached by policemen at the railway station who shared information about the militants' location. Dressed in off-duty jeans and a T-shirt, he hopped on to the waiting bus with his team of para-commandos parked outside the railway station. Seated, he went through a mental checklist, assessing who among his team he could rely on for backup in such an operation. He called Saurabh Singh Shekhawat, the indefatigable young captain in the 21 Para SF and his trusted confidant. Sengar had joined the unit to assume the role of team commander. The conversation was brief and crisp: 'Join me fast, Saurabh. I got action for you.'

Saurabh had just started his day at the training area in the team camp of the Special Forces unit when he received Sengar's call. After assembling a second team of commandos, he was trailing a few kilometres behind the bus as it arrived in the village. The plan was for Sengar and Saurabh to approach the village from different directions to surround and trap the militants.

In the waiting bus, the local informant shuffled up and down the aisle, trying to ascertain the militants' whereabouts now that the operation was in progress. He stepped off the bus, along with Sengar and a few commandos.

While Sengar kept a vigilant eye on his surroundings, the informant began to panic, doubting their decision to park, fearing it was a mistake. The house they had stopped in front of appeared unusually quiet. Just as Sengar turned to question the informant about the house, a thickset man emerged, clearly carrying a weapon. Inadvertently, the bus had stopped in front of the militants' hideout!

Mistaking Sengar in civilian clothes for someone other than an army man, the man smiled at him. In turn, Sengar swiftly drew his weapon and fired at the militant from close range, hitting him and causing him to tumble to the ground. As the enemy opened fire, Sengar's team rushed out of the bus, took positions and returned fire. A second militant appeared; Sengar fired again, taking him down. In the next moment, more ULFA cadres burst out of the house, opening fire and injuring Sengar. By then, both sides had begun to exchange a heavy volume of gunfire.

In a Jonga, a rugged four-wheel-drive utility vehicle, Saurabh was approaching the house when he heard rapid bursts of fire. He and five from his team rushed forward and gave chase. The militants, now on the run, were clearly visible in the open expanse of dry paddy fields. Saurabh stopped and raised the sight aperture of his vz.58 Czech assault rifle from

100–200 metres, aimed at a running figure and fired six rounds in single shot mode, that is, one bullet at a time. The figure fell silently before him. Some of the militants attempted to carry their fallen comrade, but they faced intense gunfire from Saurabh's men. They took cover and returned the fire.

Soon, the dry paddy fields transformed into a chaotic scene of fire, smoke, noise and haze, with everyone running and shooting. A frantic chase unfolded, resulting in a melee involving various participants: militants in combat fatigues; civilians in dhoti, kurta and denim; and army para-commandos and police personnel in battle gear. The villagers found themselves caught in the crossfire. 'Saab, bacha lo (officer, save me),' said an old man who had been injured, while a three-year-old hit by a bullet in her lower leg was transported under cover to the bus for first aid.

When Saurabh reached him, Sengar was sprawled on the ground, gripping his stomach where a bullet had pierced through, exiting from his hip. Saurabh carefully helped the injured officer into a prone position, where he lay face down, and utilized one of the seats of the Jonga vehicle as support beneath him. Seeing the officer in pain, Saurabh attempted to lighten the mood by teasing Sengar about his fitness. Sengar was perhaps one of the fittest people Saurabh had ever known, and the former never missed an opportunity to playfully remind the other about it.

'Sir, you see now, you also have a lot of fat in your body,' Saurabh remarked, playfully alluding to the accumulation of white tissue exposed by Sengar's injury.

'Oh fuck! Shut up, bloody Shekhu.' Sengar laughed, before he folded up in severe pain. 'It's hurting like hell, and you can see fat in there.'

Laughter had often kept their spirits up in grave situations, and it worked now as well. As Sengar drifted in and out of consciousness, Saurabh engaged him with light-hearted banter. He teased Sengar about leading the team, which he himself had previously commanded, from hospital. Sengar's eyes had been closed for a while, and Saurabh's face turned pale with concern for the first time. Just then, Sengar spoke, his eyes still closed. 'You can't get rid of me, you fucker.'

Opening his eyes slightly, the senior officer winked and smiled weakly. Sensing his strength was ebbing away rapidly, Saurabh said, rather forcefully, 'Don't you bloody die, sir. Don't sleep. Marnaa nahin hai abhi (You can't die yet).' In a raspy voice, Sengar responded, 'Shayad bach jaunga, agar tu time se hospital pahuncha dega toh (Perhaps I'll make it, if you manage to get me to the hospital in time).'

That day, five militants were neutralized while a few ran away. The village, however, was cleared of any presence of terrorists.

The injury from the Bongaigaon encounter had left Sengar temporarily immobilized. The hospital became his home during the healing process, and it was there that Saurabh went to see him one day.

'I see you're getting good rest, sir,' he teased his senior officer.

'So, what did you get for me, Shekhu?'

Saurabh had bought a 'Get Well Soon' card from a local Archies shop, but in his haste, he had forgotten to bring it along. 'I've brought luck, you see, sir. Waiting to have you back,' he improvised.

Sengar made an impressive recovery within a year, always grateful to Saurabh for saving his life. In the years that followed, they would once again unite in a challenging situation, strengthening their bond further.

Ambush plan unravels in Lolab Valley

The commandos from 21 Para SF, specialized in counter-insurgency, jungle warfare and guerrilla tactics, posed a significant threat to militants in the north-east region for nearly a decade. However, by the late 1990s, the momentum of the movements began to wane, as militants started surrendering in increasing numbers.[4]

The unit was in need of a new challenge, which arrived two years later. In 1999, when war broke out between India and Pakistan in Kargil, the 21 Para SF was deployed. The challenge for the unit was transitioning from jungle warfare in the north-east to operating in the high altitudes of the Himalayas. They excelled in the conflict, achieving difficult objectives, such as capturing the Neelam Post. Sengar, a war hero, led his troops in capturing Pt 5405 in Kaksar during the conflict.[5]

Even as the war was being fought in Kargil, terrorists had already been infiltrating the Kashmir Valley, hiding out in the forest, and attacking villages at will. A solution needed to be found to address and stop the increasing terrorism in the Valley. Drawing upon their recent experience in mountain warfare during the Kargil conflict and their past accomplishments in the north-east, the 21 Para SF were called in to bolster the existing counter-insurgency force, the RR, in effectively subduing the imminent threat.

The unit was deployed in the Lolab Valley in northwest Kashmir after the Kargil War ended. The Lolab Valley, once a divine masterpiece of untarnished beauty, faced the imminent risk of its claim being marred by insurgency and violence.

Allama Iqbal, the Urdu poet, eloquently depicted the enchanting oval-shaped Lolab Valley, adorned with forests of deodar and pine, with vast stretches of paddy fields in its bowl: 'Pani tere chashmon ka tarapta hua simaab (Your springs and lakes with water pulsating and quivering like quicksilver).' This picturesque landscape also embraced numerous rows of apple orchards and charming clusters of villages.

The valley was surrounded by a majestic forest of towering pines, whose thick foliage limited visibility to just about one metre. Three major foot tracks cut through the mountain at different heights. The lower track hugged the valley floor, slicing through the dense forest; above it lay a middle track which circled the belly of the hill. A third track led travellers past the top of the spur and was thus called the Upper Track.

When Major Sengar, Captain Saurabh and the 21 Para SF arrived in the valley, the possibility of an encounter with Pakistan-backed terrorist groups and local militants loomed at every turn.

One afternoon, a reliable source brought information that a number of terrorists were seen near Markul mountain and between the villages of Dardpora and Devar. The villages nestled in separate valleys within the region of Lolab and were joined by two foot tracks. One track was at a lower altitude while the other crossed the top of Markul mountain.

A 21 Para SF group set up an ambush for the terrorists. Two teams were formed – Saurabh and his group assumed control of the route at a lower elevation which led towards Devar village, while Sengar and his team of six men covered the Markul track. The entire area was thickly forested, with visibility limited to just a few metres.

By placing themselves at tactically advantageous positions, the ambush parties were able to achieve a visibility range of up to 70–80 metres along their intended route. Scouts were sent out to provide early warnings while an assault party sought cover at the identified location for an ambush, aiming to maximize their firepower and casualties. The ambush parties waited patiently at this 'killing ground' anticipating the arrival of the terrorists. Despite the lower elevation track being frequently used by villagers, there was no sign of any movement. The mountain seemed eerily silent.

Three days passed. On the fourth morning, at 0300 hours, Saurabh, positioned at the lower ambush site, suddenly heard a brief but intense exchange of gunfire pierce the silence of the night. The gunfire ceased abruptly, and once again, silence enveloped the surroundings, as if the staccato burst of a few minutes earlier had been a fleeting storm that swiftly passed over.

Just then, Saurabh's radio beeped with Sengar on the line.

'Shekhu, we've sprung the ambush.'

'Status? Over,' Saurabh asked.

'A group was here, we fired from close, but it was ... err ... too dark, can't see anything. Once day breaks, we will know. You sit tight and wait ... you might get lucky.'

In those days, night vision sights – optical devices or attachments that enhance visibility in low-light or night-time conditions – weren't yet available to the army during counter-insurgency operations. Sengar had to wait for the first light to make his move. He knew the terrorists were nearby, but it was still slightly over an hour until daybreak.

A few kilometres away, Saurabh and his boys awaited news from Sengar. At 0400 hours, on the verge of daybreak, he received a call. 'Much blood here . . . as well as a trail leading into the forest. I'm summoning the tracker dog unit, accompanied by a team from RR . . . and then I'm going to follow the trail,' he said with fervour.

Saurabh cautioned, 'Sir, because they've been hit . . . err . . . if someone is injured, they will assume a defensive position and stay vigilant . . . and we are moving with the tracker dog who will run and generate sound . . . making the terrorists alert to our movement . . . be careful.'

Sengar reassured him that he had taken all necessary precautions. 'Under control . . . yes . . .' came the crisp response.

Waiting, the team at Devar grew increasingly impatient. Sudden changes in operations were not uncommon, reflecting their inherent unpredictability. On duty, prolonged periods of silence often gave way to sporadic gunfire. While most soldiers believed that accurate firing could determine the outcome, whether it be good or bad, of an operation, Saurabh had a different perspective. To him, a fight was won or lost in those moments of silence. He believed that those who maintained the element of surprise would ultimately prevail.

Saurabh glanced at his watch; just over an hour had passed since he and Sengar had spoken on the radio. A few minutes later, the grating sound of automatic gunfire erupted. Intense bursts of gunshots were followed by loud blasts, occasionally interrupted by a brief pause, as if the gun battle had been called off. Immediately after, the automatics would open up again. As the barrage intensified, Saurabh tried reaching Sengar but he received no response.

The radio beeped after a few minutes, and it was Sengar on the line. He sounded calm and reassuring, as always, though he spoke at a deliberate pace. 'I've been hit ... may not survive now. But ... you come here fast ... there are a lot of terrorists here. Don't let them escape,' he said, trailing off.

Saurabh called back and this time he got Commando Duttaram Ghale on the line. 'Sir ... Sengar saab ko goliyan lagi hain, sab ki halat kharab hai (Sengar sir has been shot, and we're all in a bad state),' Ghale said.

'Baaki log kahaan hain (Where is the remainder of the deployed team?)?' Saurabh asked.

'Sir, kyunki tracker dog bhaga thaa, toh hum uske peeche peeche bhage. Mera unke saath koi contact nahin hai (The tracker dog ran ahead, and we chased after it. I have no contact with the others.),' Ghale said.

There was only one AN/PRC-25 portable radio set amongst the team. Establishing communication with Sengar's other team members would have been impossible for Saurabh. Therefore, he quickly decided to move reinforcement to the downed team. Racing to the nearby Devar post, he offloaded the heavy equipment to ensure the ascent went smoothly. He retained his personal weapon,[6] a vz.58 assault rifle, a PK machine gun, grenades and a rocket launcher. Each member of the team carried one or two high-explosive rounds for the rocket launcher. Alongside his buddy (assistant) Narayan Limbu and the team, Saurabh ran into the paddy fields, maintaining a distance from the upslope of the hill as they advanced to avoid being ambushed by terrorists alerted to the arrival of reinforcements.

The commandos were tough runners, having trained through the year in the mountains for such situations, and quickly covered around 5 kilometres. While the sound of intermittent gunfire pushed him to cover the distance faster, Saurabh's mind was a whirlwind of thoughts. 'How do I save Major Sengar... he was hit again... how long can Ghale fight alone?' A sense of déjà vu enveloped him as he considered how to swiftly evacuate Sengar again. Would he be lucky a second time round?

Saurabh briefly paused his run, straining his ears to listen, but the sound was muffled by the dense forest. He needed to find Ghale quickly, and tried to determine the direction the gunfire was coming from.

Evacuating the severely wounded major

Although descending from a long line of daredevil Gorkha soldiers who had served in the Indian Army, Duttaram Ghale had chosen to join the 21 Para SF. Unperturbed by that morning's events, the dogged fighter had single-handedly delayed the terrorists after Sengar went down. He pulled the latter to safety behind a fallen pine tree, using it as a shield to hold off the terrorists, even as he coordinated with Saurabh over the radio.

The radio line crackled and hissed, causing communication to be unclear. 'Saab, mera mission khatam hone wala hai (Sir, my mission is about to end),' Ghale said, attempting to convey that he was running out of ammunition. Saurabh asked him to repeat himself a few times and share his location. He then told Ghale to fire a double tap in quick succession. A double tap is a distinctive set of individual shots fired in single-shot mode during heavy firing to help distinguish its unique sound.

Crack! Crack!

Listening carefully for the shots, Saurabh now realized that Ghale was on a lower slope.

The terrorists, anticipating that the army would come searching for them from below, had strategically occupied the higher ground.

A heavy automatic fire on Ghale's position helped Saurabh estimate the terrorists' position to be about 200–250 metres away. He decided to bring out the big guns and signalled to the rocket launcher operator. 'Okay. Let's go... HE rounds!' he said, ordering high-explosive ammunition fire. The soldier hauled the 84 mm Carl Gustaf portable launcher over his right shoulder and fired. The airbursts generated a devastating fireball that shook the terrorists into silence.

Saurabh seized this opportunity to climb a few metres and managed to reach Ghale's position. There he found a bloodied and wounded Sengar in critical condition.

In previous operations, Saurabh and Sengar would often exchange banter about the latter's injuries. However, this time, Sengar's face had turned pale, his eyes were vacant and he was deteriorating rapidly. An urgent request for emergency evacuation by helicopter was made.

As they waited, Saurabh took a closer look at the major's wounds. Five bullets had torn through the lower half of Sengar's body. His thighs had been shredded, waist and hip bone had been split open and, to make matters worse, a rocket-propelled grenade splinter had pierced his chest.

One of the men administered a dexamethasone injection

to alleviate the pain. Despite the application of the first field dressing and applying every piece of their own clothing that they could spare, the men were unable to control the bleeding. Recognizing that Sengar was closer to death than ever before, Saurabh silently offered a prayer.

Just then, a faint whirring sound in the skies was heard. As it grew louder, the men realized that it was the rescue helicopter. At this point, no one knew if Sengar would survive. The men who had seen his injuries doubted he would. However, the mere sight of the helicopter landing injected a renewed sense of hope.

Despite his wounds, Sengar's focus remained on the mission. He instructed Saurabh to swiftly take down the threat. His concern was not for his own life, but for the fight at hand.

Sengar was stretchered into the helicopter and, along with his buddy, paratrooper Rajesh, evacuated within minutes. As he watched them fly away, Saurabh wondered if he would ever see the Major again.

The young officer paused, closed his eyes briefly, took a deep breath and turned to Ghale. They exchanged silent nods, understanding that there was work to be done.

Confronting the terrorist in the pine forests

Under cover, Saurabh and Limbu cautiously ascended the slope. At once, they spotted a pool of blood on a tiny patch of grass and the body of a RR jawan from the unit Sengar had called in as backup. They also heard the muffled sounds of whimpering from behind the bushes.

They came upon a distressed young soldier, who seemed

shocked and disoriented, covered in blood. On seeing Saurabh, he lashed out. 'Bahut bura ho gaya, sir . . . bahut bura! (It has been very bad sir, very bad!),' he shouted, before breaking down.

Attempting to console him, Saurabh ordered the soldier back to base. Unfortunately, he refused to budge, even when rebuked, as Saurabh tried to comfort him, the soldier looked up with forlorn eyes. Slowly, he got to his feet, wiped the tears off his face and picked up his rifle. Suddenly, he raised a frenzied war cry – 'Raja Ram Chandra ki Jai!' – and fired his rifle in the direction of the terrorists. Saurabh swiftly pounced on the soldier, pushing him down to the ground, the young officer held him by the shoulder and looked him in the eye. 'All will be fine, we will take out whoever is hiding there but you need to leave,' Saurabh reiterated.

The soldier refused. 'Saab, mera buddy chala gaya, mere ko bhi jaane do, ya main nahin ya ye banda nahin (Sir, my buddy is gone, let me go too, it's either this guy or me here).' He appeared disoriented and exhausted. The harsh realities of intense operations in 1990s in Kashmir often led to extreme shock and disorientation among the toughest soldiers, impairing their ability to think clearly. Many returned from tenures in the Valley carrying visible scars of trauma.

Saurabh attempted to reason with him. 'Get up . . . get up . . . you're a brave man. Brave man you are! You fought well. Don't cry, everything will be allright, okay?' This time, the soldier promptly obeyed the order to leave, seeking cover from the intermittent enemy gunfire coming from the mountain-top. No sooner had he departed, a stubbled man, clad in a blue tracksuit, emerged from the bushes. Annoyed and visibly perturbed, Saurabh confronted him, weapon in hand, demanding, 'Who are you now?'

'I am Sparrow... don't fire!' the man protested. Nicknamed Sparrow (the call sign), he was a Signal Officer – responsible for managing communication systems in the brigade – who had accompanied the tracker dog team. He was ordered to take charge of the RR soldier and accompany him down the hill.

Saurabh's focus was now on the task ahead. Limbu had spotted the barrel of Sengar's rifle about 20 yards away, abandoned from the earlier encounter. On Saurabh's instructions, he surreptitiously crawled towards the spot. Just as he reached for the rifle, a burst of gunfire erupted from the bushes, causing him to topple down the mountain slope. His fall saved him from the firing line of the second burst that followed.

In the meanwhile, Saurabh had taken cover behind a pine tree and was biding his time to take a crack at the enemy. A few rapid bursts struck the trees, causing bits of bark to fly as bullets ricocheted. Saurabh crawled and positioned himself behind a large tree stump, finding both cover and a favourable firing position.

He heard rustling bushes to his left. Behind a tree, he found the dog and trainer crouched in hiding. 'Now, what are you doing here?' Saurabh exclaimed, impatient with these unexpected encounters.

'Saab, there is heavy fire coming down on us. If I expose myself and the dog, we will die,' the trainer said.

Glancing in the direction of fire, he instructed the dog trainer to stay behind the tree blocking the incoming fire and proceed straight down. 'Seedhey jaana (Go straight). Don't veer.' Following Saurabh's directive, they found their way down unharmed.

Hopeful that there would be no more surprises, Saurabh estimated the fire came from about 10–12 metres away. The forest cover made it appear further away.

As he crawled forward silently, he heard Limbu from his hiding spot. NL, as Saurabh called him, was a young man from eastern Nepal who knew neither fear nor local expletives when he had joined the unit. Having served and lived among Rajputs, Sikhs, Jats, Dogras and Marathas in the 21 Para SF, NL now knew the choicest of expletives, which he unleashed at the terrorists.

Saurabh tossed a pine cone at NL. Once he had his attention, the officer signalled for him to take cover and stay silent, so as not to reveal his position. Suddenly, a terrorist, annoyed by NL's abuse, spoke. 'Gaali kyun de raha hai (Why are you abusing me)?'

Saurabh, straining to detect even the slightest sound, instantly realized that he had crawled to a more advantageous position than the terrorists and was positioned approximately six or seven metres above them.

'Mujahid bhai, kahan ke rehne wala hai (Fighter, where are you from)?' he asked.

Surprised by the softer tone, the terrorist responded to the friendlier voice. 'Janab, Pakistani hoon (Sir, I am Pakistani).'

Saurabh was impressed by how calm the man sounded after hours of gun battle. Matter-of-factly, he said, 'Mujahid, you are now surrounded. It's best you surrender or you will be killed.'

Recognizing that he was speaking to someone in a position of authority, the terrorist responded politely. 'Janaab, ye badshahon ki ladhai hai, aap aur hum keval pyade hain. Aap apna kaam karo, main apna kaam kar raha hu (Sir, we are serving the orders of Allah. This is a war between kings; you and I are just foot soldiers. You do your duty and I will do mine).' Respectfully, he declined the offer to surrender.

Saurabh changed his tack, unclipping a grenade and lobbying it towards the terrorist. The grenade rolled down the slope, past the terrorist, and exploded. Saurabh threw another one, whose blast tore through the vegetation. He heard a stifled cry from behind the bushes and, after a short while, asked, 'Mujahid bhai . . . are you hurt? Did that get you?'

'Janaab, I took a bullet last night and injured my calf and now . . . yeh grenade jo phenka, mere haath mein lagi (this grenade has damaged my hand).'

'Tu yahaan naahaq maara jayega (You will be killed unnecessarily). Toh ab kahaan jayega (Where will you go now), Mujahid bhai?'

A pause in dialogue ensued, and no one moved on either side. Then Saurabh shouted, 'Hathiyar daal do (Throw down your weapons). Surrender!'

'Humey toh Allah ka hukum hai, hum to unka kaam karte hain (We have Allah's blessings, we will continue his work).' In that moment, the terrorist's voice hardened.

The two warriors faced each other, driven not by malice but by conflicting ideologies and religious beliefs. Each sought the other's death. In their ongoing 'search and kill' encounter, characterized by alternating moments of chatter and firing, neither side showed any sign of weakness. Despite provocations, both had maintained a civil front.

Saurabh pondered over the Mujahid's choice to remain in the same position. Was it due to an inability to move further, or was he simply biding his time for the right moment? He waited for the terrorist to fire, while the latter seemed to anticipate Saurabh making a mistake.

The terrorist knew Saurabh was an officer, whose death would bring him glory. This was his chance to create a lasting legacy.

A shot in the dark

Paratrooper Manju Shah, who was Saurabh's radio operator, had been listening to the conversation. Following Saurabh's voice through the thick undergrowth, he crawled forward. Saurabh, whose eyes were trained on the terrorists in hiding, caught a glimpse of Shah's shadowy face in the darkness. He wasn't the only one who could see Shah.

All of a sudden, there was firing. The radio operator fell backwards and rolled into the bushes.

'Manju Shah . . . are you hit?' Saurabh shouted. He received no response, so he asked again. 'Are you hit?'

'Nahin saab (No, sir), bullet hit my radio set.' Shah now lay on the ground beside him.

The AN/PRC-25 portable radio set had been Shah's saviour. The bullet fired at him bounced off his radio set instead of penetrating his back or shoulder. Crouched in the same dark location for an extended period, Saurabh's eyes had become accustomed to the surroundings. Focusing intently on the spot from which the terrorist had fired at Shah, he eventually discerned an obscure oval shape amid the bushes. He took aim and fired a single shot.

The figure disappeared.

Saurabh fired another shot and watched it cause the bushes to shake.

A deafening silence followed.

The Indian officer strained to see any signs of movement as he scanned the area. It was so silent that even the slightest movement of an insect could be heard.

'Mujahid bhai? . . . Mujahid?' he said, his eyes fixed on the spot and his hands firmly on the trigger of his cocked rifle.

No answer. Saurabh crawled forward at a measured pace. The encounter lasted longer than he had expected. The crafty terrorist had remained hidden, while NL patiently awaited his moment, poised to strike.

NL looked at his watch. A few minutes had passed since Saurabh had fired the shot. On the latter's instructions, NL began crawling upwards from his position below. Reaching the spot, he shouted, 'Saab, he's dead . . . he's hit.'

Saurabh and NL looked down upon the fallen man, a tall and lean figure with a scruffy beard. Later, intelligence sources would identify the man as Abu Khalid, the local commander of a prominent terrorist outfit. With his skilful use of words, bullets and camouflage, Khalid had managed to hold off the Indian forces for hours. Saurabh couldn't help but be impressed by his courageous stand. He contemplated the terrorist's origins. Based on his language and appearance, Saurabh speculated that he was of Punjabi, Muslim Rajput or Jat descent, potentially sharing the same ancestral heritage as him. Unlike him, however, the terrorist had been influenced by false narratives and manipulated by handlers who promoted radicalized, militarized Islam.

On closer inspection, they learnt that Khalid had an

injured leg and, knowing that he wouldn't be able to keep pace with the rest, had volunteered to take on the Indian Army patrol while allowing his team to escape. Terrorists typically carry three to four magazines, but this young insurgent had fourteen magazines of ammunition next to him when he was killed. He had stashed extra magazines with him to prolong the engagement with the army while his comrades made their escape. He had had the chance to surrender but chose not to. That's when Saurabh realized the terrorist had been alone, waging the perfect guerrilla warfare. Saurabh had ended the battle with a bullet between his eyebrows. It had been the Indian officer's best headshot as a marksman.

Ahead of them was a scene of carnage. The body of Kishan Singh, the scout from Sengar's ambush party, was discovered alongside that of Shehzada, Sengar's local guide, whose body was riddled with bullets. They had died before Saurabh arrived on the scene. Three Indian soldiers had been lost and Sengar was battling for his life in a Srinagar military hospital.

Over the next ten days, operations were launched to track down the rest of Khalid's men in the jungle. Within a week, they were rounded up and taken down. Some died, others were captured due to injuries sustained. The death of their young division commander had dealt a significant blow to their plans.

The 'Get Well Soon' card finally arrives

There is a popular belief that those who make it to 92 Base Hospital in Srinagar's Badami Bagh Cantonment alive are unlikely to succumb to their injuries.

In the early 1990s, the hospital witnessed an influx of

casualties from intense counterterrorist operations. Every day, doctors treated patients with bullet injuries, explosion wounds and combat-related damage. Over time, the hospital acquired a reputation for successfully treating critical cases. Sengar's condition fell in that category.

Three weeks after the encounter, Saurabh walked into the hospital carrying a bouquet and an envelope. Sengar smiled as he entered his room. Wrapped in plaster, he lay in bed awaiting transfer to the Army Hospital in New Delhi. The extent of Sengar's injuries was outrageous – dislocated body parts, shattered bones, bullet wounds in his chest and abdomen, and damage to his hip bone. The doctors started by fixing his hip bone, warning him that one leg would be shorter than the other. They believed he would never walk again, but Sengar would defy their expectations in the years to come.

Saurabh set down the flowers and handed the 'Get Well Soon' card to Sengar, whose eyes immediately went to the date. It was from two years earlier – 1997, when Sengar had sustained previous injuries. On the card, Saurabh had written, 'Thanks for getting injured again, you saved my card.'

'Date rehne deta (You could have skipped mentioning the date),' Sengar said with a wink.

'Sir, it was meant for today. Destiny. That's why I never gave it to you earlier,' said the younger officer, smiling. Jokingly, he added, 'You have more bullets than bones in your body. Should I call you metal man? No, wait, you're Superman ... even better, SuperSengar!'

'Thanks, mate,' the injured warrior said, his eyes gleaming. 'I reached here alive. That's all I had to do ...'

Saurabh shook his head and indicated the ghastly wounds on Sengar's lower body. 'Sir, please get yourself bulletproof underwear since you always attract bullets to that area,' he said teasingly.

No one had spoken to Sengar in such a manner, and for the first time since being wounded, he burst into hearty laughter and even playfully threw a mock punch at Saurabh.

Despite being shot, Sengar continued to provide calm leadership and able direction till he was evacuated. He had defied the odds and stayed alive, but he was right about one thing – he owed his survival to the unwavering support of his men. Ghale played a crucial role in bravely confronting the terrorists. Saurabh's swift action ensured the helicopter arrived in time. His buddy Paratrooper Rajesh stood guard by his side during the evacuation. The doctors had worked tirelessly to revive him, and Sengar's determination gave them hope. He had triumphed over death once more. And as he recovered, he received the news that Saurabh had eliminated the terrorist leader and their entire group.

Sengar had cheated death yet again, living to fight another day.

Postscript

The soldiers in this story embodied the warrior's code of esprit de corps (loyalty and pride shared by a group) and unwavering resilience, even in the face of intense combat, living up to the Chetwode Motto at the Indian Military Academy. 'The safety, honour, and welfare of your country

Major Deependra Singh Sengar (pic taken when he was a captain)

come first, always and every time. The honour, welfare, and comfort of the men you command come next. Your own ease, comfort, and safety come last, always and every time.'[7] That day, leaders on both sides – Sengar and Saurabh on one side, and Abu Khalid on the other – had displayed exceptional moral courage in the face of death.

What became of these brave men?

Major Deependra Singh Sengar was awarded the Sena Medal for his gallantry in Assam, Sengar later faced death in the Lolab operation. Despite several operations, the medical prognosis for his condition was disheartening. He was told that he would likely never regain the ability to walk without crutches. This devastating news meant that he would be unable to return to the Special Forces. Undeterred, he resigned from the Indian Army, successfully prepared for the Common Admission Test (CAT) from a hospital bed and went on to graduate from the Indian Institute of Management, Ahmedabad.

Miraculously, Sengar survived the ordeal and went on to become a global leader in the corporate world. Today, he lives in the United States with his family.

He calls himself a soldier of destiny.

Abu Khalid was a young terrorist dedicated to a misplaced cause, who displayed an unusual calm and civility in the face of death. As a leader, he allowed his men to escape while he dug his heels in and fought well. He chose death over surrender. (This isn't an attempt to glorify the enemy but rather to shed light on the unwavering determination of a terrorist.)

Brigadier Saurabh Singh Shekhawat remained with his unit throughout the tenure in Kashmir. He has achieved a remarkable record of always bringing back his men alive from serious combat, even in the face of daunting challenges and high-risk missions, including those considered suicidal.

Saurabh is a remarkable individual with a passion for fitness and adventure. He has conquered twenty-one mountain peaks, including Mount Everest three times. He is also an avid ultramarathon runner and has completed the challenging 72 kilometres Khardung La Challenge in Ladakh. He entered the *Guinness Book of World Records* by running the Pangong Frozen Lake Marathon in February 2023. He is a passionate horseman, which he pursues as both a hobby and a family legacy.

Saurabh is one of the most decorated serving officers of the Indian Army, and has been awarded the Kirti Chakra,

Shaurya Chakra, Sena Medal and Vishisht Seva Medal for his various acts of valour.

In addition to his remarkable bravery, Saurabh is known as the Indian Army's quintessential Iceman – a calm and unflappable crisis man.

Narayan Limbu took premature retirement and settled in his home country.

Duttaram Ghale and **Rajesh Balhoria** went on to become JCOs.

This story is based on the account of Brigadier Saurabh Singh Shekhawat.

9

The Militant and the Major
Finding Nizamu

'Any news of Nizamuddin?'

The CO of the RR battalion entered the officers' mess, looking irritated. His terse tone startled the group of officers huddled in a semicircle around the bukhari, a traditional heating stove. The hot cylindrical drum provided much-needed warmth during the harsh winter of the Kashmir Valley.

It was the year 2006 and militants from across the LOC were trying to regain authority, fighting the state police and scrapping with the Indian Army. Six years earlier, when war broke out between Indian and Pakistani armies at the Kargil border, these militants sneaked in through the mountains, pushed by their handlers in Pakistan into the simmering unrest in the Kashmir Valley.

The RR, a component of the counter-insurgency force established by the army in the early 1990s, had earned its spurs in the challenging and dangerous valley. Here, young officers and soldiers confronted insurgents head-on, earning their reputation. By the end of their tenures, however, they were a weary bunch, longing for a 'peace station' away from the conflict.

For the CO, leading an infantry battalion of about twelve-hundred combat-trained men, Nizamuddin was a prized catch who had eluded them. Rumoured to be the brain behind the rising local militancy movement, officers and men often asked, 'Has anybody seen him?' Each time, the question was met with silence.

Reportedly, Nizamuddin had been spotted on two separate occasions a month earlier, but these sightings had proved to be red herrings. There had been no news about him since then. His appearance and whereabouts remained a mystery to everyone.

As the days went by, the stories surrounding Nizamuddin grew bigger and more intriguing. There were tales of him helping a local family with money and another about him being idolized by a village. He was turning into a legend, and the CO knew that a Robin Hood figure could become an even bigger headache. 'Sir, we are trying to find him . . .' muttered one of the officers, without much conviction. The elusive tales only heightened the sense of suspense and anticipation among them.

A younger officer sidled up to Major Mohan Sundaresan Kumar and said, 'If we catch him, it'll be a big one, sir.'

That night, Mohan, the company commander, was driving back to the Delta Company HQ in his military green Maruti

Gypsy when he noticed the figure of a man draped in a pheran (long robe worn by Kashmiris) next to the gate.

Mohan remained unperturbed, as their local sources often crept in at night to share significant updates or alarming news.

'How are you, Shabbir? What happened . . . you're here so late? Khairiyat (All well)?' he asked.

Shabbir, a local cab driver, was often privy to revelatory conversations between passengers who spoke freely in the back seat of his car. Some of his passengers would boastfully share stories, while others' disclosures spurred Shabbir's gumshoe spirit.

'Haan sir, sab theek hai (Yes sir, all is well). But one thing . . . I want to tell you.' Shabbir stepped closer as Mohan got out of the car. 'I have important information.'

He spoke in a hushed voice to avoid being heard by Mohan's radio operator. The major was stunned when he heard the news.

'How do you know? Are you sure?' he asked urgently.

While Shabbir usually supplied accurate information, for the past month he hadn't come up with any substantial leads. For a moment, Mohan wondered if this tip-off was a manoeuvre to stay relevant.

'Where do we have to go?' he asked.

Shabbir pointed to the area across the river, which came under another battalion's purview. The driver quickly added that while he was aware of that, he had come to Mohan because he trusted him.

There were three villages across the river. Two of them stood next to the headquarters of the local army unit. The third village, a couple of kilometres away from the army unit, housed a government school.

Shabbir mentioned a house in that area.

'Isn't that a teacher's house?' Mohan queried.

'Yes, she is with a local government school.'

'We have to get to her house and search, is that it?'

'No sir. We don't have to search... because I know. You see ... I am not wrong,' Shabbir said with conviction.

The hunt ends

After exchanging their uniforms for civilian clothes, Mohan and his men were soon on their way in taxis and non-military vehicles.

The young company commander was well aware of the risk this operation entailed.

He was heading into an area controlled by another battalion where he could be stopped and detained or, even worse, ambushed and killed by friendly fire if he was mistaken for a militant. He and his men had grown beards to blend in with the local population. Travelling in civilian clothes and using taxis and minivans while carrying weapons made them vulnerable. Thoughts swirled inside his head, foremost being what lay ahead.

Shabbir guided them beyond the main road, leading them to the former government health office that now served as the school building. Passing by the stone and mud huts, he pointed towards a house situated at the village's periphery, overlooking the river. Crafted from Deodar wood, this single-storey house featured ample windows on every side, a sloping roof designed to shed snow during winter, and a sturdy front door. One of Mohan's men rapped on the door. Covert operatives stood surrounding the house, their rifles in position and ready to fire, the quiet clicking of their cocking handles filled the air with edgy anticipation.

The front door opened, and a woman stepped out. She was pale, with an earnest expression that shone in the night. A former militant, who now acted as a guide for Mohan's team having surrendered to the authorities, stepped forward. 'We have come from the other side,' he said reassuringly. The woman nodded and shut the door.

On the doorstep, Mohan's man stood exposed and vulnerable, and each passing minute felt like hours. The men surrounding the house waited too. Mohan held his breath, feeling his pulse racing with anticipation. His thoughts went to the handbook every officer was given. It had pictures of militants of all types – dreaded, wanted, fledgling, etc. The pictures were old, and the faces had changed, rendering them inaccurate, but officers still carried the book. Mohan wondered if he had ever come across the elusive Nizamuddin within its pages.

Just then, the front door opened with a slight creak and was left ajar. Mohan's men aimed their rifles at it. In stepped a man; Mohan could see he was in his early 30s, had large eyes and a well-kept beard that added some gravitas to his appearance.

The former militant on the doorstep nodded to confirm the man's identity. The man came forward and now stood before Mohan and his men. The team stood in stunned silence, recognizing the enormity of the moment. They had waited for months and now, finally, he was here before them, unarmed, smiling and with his arms raised. There were no signs of resistance from the dreaded militant, who recognized

Mohan. 'Major saab, chalein (Major, shall we go)?' he said, his diction crisp, unlike that of the other militants they had come across.

He was bundled into a waiting Sumo and driven off to camp. Mohan wore a quiet smile.

He finally had Nizamuddin!

A militant returns home

The next morning, as the sun shone brightly, Mohan sat in the shade with a cup of tea and prepared a report for his CO. He had planned to radio in Nizamuddin's surrender. While shuffling through the information on the man in their custody to add to the report, he received a message. Nizamuddin wanted to meet the company commander.

'What could be the reason? Bring him,' Mohan instructed.

Following his arrest on the previous night, Nizamuddin had been placed in the detention cell. Now, sitting before Mohan, he appeared willing to chat. He began by saying, 'We were all students when this thing... this movement... found momentum. Kids were joining, inspired by what they heard. I am padha-likha (educated), you know... an MCom.'

'So, why did you join them?'

'I have a commerce degree. I look after their accounts.'

'Why did you surrender?'

'I have information to share, Sahib. Weapons...'

'Tell me... do you know where?'

'Come, sir. I will take you.'

Mohan wondered how many weapons there could be and what types they were. He called the company's havildar major. 'Quick... get the boys together,' he ordered. That afternoon,

Nizamuddin's information helped them seize a few rifles and hand grenades.

As they headed back from the operation, Mohan put on his company commander thinking hat. A daring thought entered his mind. 'Should I hold the news of Nizamuddin's capture for a while?' he pondered.

In the Valley, company commanders of battalions pursued every opportunity to recover weapons and capture militants. It was a key responsibility for them, as the army's success depended on the efforts of the ground forces. Withholding news of his arrest and having Nizamuddin in their custody could provide them with valuable leads on weapons and militant locations.

Mohan decided to trust his instincts, and the events that unfolded over the next four days confirmed that he was right. Nizamuddin's leads proved to be very valuable. They ventured out twice in search of weapons and returned with success on both occasions, narrowly missing the capture of militants.

Nizamuddin quickly formed a bond with the men in the battalion; they played carrom and cards together and even washed their clothes side by side. Mohan also enjoyed the surrendered militant's company and occasionally invited Nizamu – as he was now known – to have lunch with him.

Outside the company langar (cookhouse), over a bowl of chicken, rice and dal, the major and one-time militant had a deep conversation.

'So, tell me. You're an educated guy ... padha-likha ... what do you think about Kashmir?' Mohan asked.

'Sir, people, and the youth especially, are confused and scared of everyone, be it militants, fauj (army), police ... it's the politicians, they play a good game.'

'And which side are you?'

Nizamu grinned, revealing his tobacco-stained teeth, a testament to the years he spent living on the road. 'I have been there long enough, saab. A "hardcore militant", as you like to say, saab.' He disclosed his affiliation with the Hizbul Mujahideen (HM), an Islamist separatist organization that stands for the integration of Jammu & Kashmir with Pakistan.

Mohan, who had been mulling over the night of his arrest, said, 'Why didn't you put up a fight?'

'Huh . . . let me tell you the story. I was young . . . educated, capable bhi tha (also). My mind, however, was drawn to the cause, to what they (militants) said . . . so I decided to join them. Main dus saal raha (I stayed ten years), sir. Ten years. No one else there was educated . . . I knew accounts, used my skills and quickly rose up in the ranks. Biggest thing is, I won their trust.'

Nizamu sighed deeply and wiped his eyes, shrugging his shoulders.

'That doesn't answer my question, Nizamu,'

The major gazed into Nizamu's eyes, finding little emotion or vulnerability. Mohan had witnessed enough militancy to remain unaffected by the tale.

'Yes . . . sir, the information that came to you . . . was planted by HM, you know.'

'Why would they give information about their own guy? Just like that?' Mohan said as he smirked.

'No, they won't . . .' Nizamu smiled weakly now. 'But see sir, I am not with HM . . . I have left them.'

Nizamu tells his tale

Taking a deep breath, Nizamu revealed his entire story.

After his rapid rise in the HM ranks, he quickly took charge of all their bank accounts and gained the trust of the top leaders. To outsiders, he appeared as a towering force, instilling fear as a top militant. Villagers would cower on seeing him and policemen privately hoped they didn't encounter him in an alley.

He was the rising star of Hizbul.

One day, he was in the area with his men, roaming with an insouciance of a dreaded militant, when he happened to spot a pretty woman. She wasn't the first he had seen in the area, but she was the only one who appeared unruffled by his presence. The next time he saw her, he was with his group, carrying his guns and intimidating the frightened villagers.

She refused to even acknowledge him.

Drawn to her calm resolve and confident swagger, he found himself trying to draw her attention. His friends told him her name was Mehrunissa and she taught in a local school. For a change, the feared Nizamu had to summon courage to sidle up to her one day and reveal that he had a master's degree. 'I am educated, just like you,' he said, feeling embarrassed to resort to such a lame line after all those years just to impress her.

Mehrunissa had smiled nonchalantly and said that his education was useless since he had picked up a gun. Not one to give up easily, Nizamu persisted, and soon, the two would meet and talk whenever he visited her village. Once, when he was unwell, he even stayed over at her house. During the course of their friendship, they eventually fell in love.

Mehrunissa told him that he would have to give up his gun if they were to live together.

'Give up means ... give up forever?' he had asked her.

'Yes, give up militancy and surrender.' She had shown him the way out.

One night, he managed to slip away from the HM camp. Being a senior operative, no one suspected him of leaving. The next morning, Mehrunissa heard a knock on her door – it was Nizamu, who had given up the gun for good.

For militants looking to abandon their cause, there was always the looming threat of retribution from former comrades. They needed a safe place to lie low until they could trust an officer to surrender to.

Nizamu, wearing a disarming smile, stood at Mehrunissa's doorstep, and the young teacher quietly embraced her lover, welcoming him home.

'When the HM learnt what I had done, they wanted me to be picked up. That's how you got my location so easily. Your khabri (source), who tipped you off, has connections with the HM. It was all prearranged, you see,' Nizamu explained.

Mohan continued to stare, taking long puffs on his cigarette, which made Nizamu uncomfortable. He continued, 'I am fed up. I don't think militancy makes sense.

'I turned to you; no one else would have heard me out. I had faith that your informant would brief you about the situation. Surrender kiya maine (I surrendered).'

His explanation was met with silence.

After a pause, the former militant said, 'I have a question, sir. One day you may have to kill me. Maar doge (Will you kill me)? Your bosses don't know I am with you. You're taking a huge risk.' The bearded renegade threw back his head and guffawed.

That evening, Mohan called Nizamu for a drink and sat down with a bottle of Old Monk rum from his duty-free ration. Nizamu continued with their earlier conversation. 'I wanted to test my luck when I decided to surrender. How long would I live? Someone or the other is likely to shoot me . . . so why not surrender to someone and see?'

He paused. 'You know . . . I thought of moving to New Delhi. Settle down, kuch kaam-waam karo (get a job) and then maybe go abroad . . .' He trailed off.

They were halfway down the first drink when the radio operator ran in, struggling to catch his breath. He brought bad news. The battalion CO was on his way. The sentry followed with an update. The boss was at the gate now. Mohan contemplated whether he had made the right decision by probing for more information, or if it was a greater mistake to keep his boss in the dark.

Before Nizamu could be sent back to his cell, the CO walked in. Clearly, he had caught wind of something and wanted to confront Mohan.

Taken by surprise, Nizamu quickly composed himself and addressed the visitor. 'Hello saab,' he said with a grin and extended his hand. The CO shook it, and then took Mohan aside. 'You didn't inform me. Why?' he asked, looking him in the eye.

Mohan stumbled, trying to convince the CO that Nizamuddin had provided intelligence. 'We have got weapons, sir . . . I was about to tell you all that. We expect some more information from him.'

'Mohan, you're crazy. I am disappointed you had more faith in that guy than you have in me!' The CO's contempt was evident, and Mohan braced himself for an uncomfortable order.

'Sir, would you like to sit down? We can have food . . .' he said in an attempt to pacify his boss.

The CO refused with an excuse about another operation that required his attention. He walked away briskly, but halfway to his car, he stopped and turned to Mohan. 'Bump him off,' he ordered.

Jolted, Mohan tried to reason with him. 'Sir, he has surrendered, given himself over to us . . . he has led us to weapons. Says he will get us more people . . . we have to give it some time.'

The boss had the last word, 'Bump him off, I said. Period,' in an ominous, incontestable tone.

Following orders from the top

An infantry battalion company commander operating in a conflict zone has trusted hatchet men who are unflinching in carrying out the most gruesome tasks. At times, these tasks seem at odds with reality. But then, the hatchet man's job isn't to seek the truth. Instead, it is to eliminate it.

Havildar Radhe Shyam, a Jat from Haryana, came from ruthless stock, and was raised to protect his people, land and izzat (honour). While the army had tamed his instincts to an extent, his unrepentant nature and ability to handle tough jobs made him the company commander's favourite hatchet man.

After being reprimanded for not gathering sufficient evidence during an earlier militant encounter, Shyam ensured he followed protocol when Mohan gave him orders. Shyam took on a terrorist bare-handed and killed him brutally. Unable to carry the body back from a remote mountain area near the LOC, he collected a few knocked-out teeth and cut off the terrorist's ears as evidence to show the officers at HQ that he had completed the task.

When Mohan received the CO's orders regarding Nizamu, he knew who to call. 'Radhe,' he bellowed.

On receiving his orders, Shyam said nothing. He simply stared at his company commander for a few minutes and then left.

The next morning, Mohan was sipping tea when his buddy (assistant) said, 'Sir, Nizamu wants to see you.' Nizamuddin the militant had now become Nizamu to everyone in the company.

The company men knew what was about to happen since the moment the CO had shown up. Mohan had asked them to be ready the next morning.

When Nizamu arrived, he pretended to be calm though he was seasoned enough to fathom what was in store for him. Mohan sensed he was anxious to share a secret before he departed. 'You know, there are two things I want to tell you,' he began. 'I will give you a pen drive that has information stored in it. About the money, where it's hidden.'

'Pen drive...? Why are you giving it to me?' Mohan wanted to know.

'Yes sir. I've hidden the money. I know these guys here are going to kill me, I know it, sir. Isn't that so?' Nizamu smiled sadly.

Mohan kept a straight face and repeated his rehearsed lines. 'No, well . . . there's information that there are some suspicious people in the forest area a little ahead. I don't have the time, but the boys will take you. You can show them the way.'

Nizamu shrugged and turned away.

There were ten men, all armed and ready. Mohan watched as they took Nizamu away, disappearing down the slope in front of the company area. They emerged further ahead, following the path that led to higher ground. Their destination was a place that held many secrets in the valley – the Jhelum river.

Mohan had fifteen minutes to make a decision. He had to act immediately or Nizamu could wind up in the river.

He bolted out of the company area and ran faster than he had as a cadet in the academy. It brought back memories of a time when Mohan had tried to save his mate from relegation by secretly running the 5-kilometre Battle Physical Efficiency Test for him at the academy. Unfortunately, he was caught and punished, and his friend was demoted.

Mohan couldn't help his friend then. He wouldn't fail another today.

Sprinting through the jungle and leaping over the forested trail, Mohan reached the location. He saw Nizamu on his knees and Shyam holding a gun to his head. Two men stood over the duo, as backup for Shyam, while the rest of the group guarded the incoming foot tracks and kept a watch over the area around them. Nizamu was giving Shyam advice on how to shoot. 'Here, here . . .' He pointed to his temple. A ruthless man, Shyam appeared sickened by the task, but it was the boss's order.

'Stop, stop!' Mohan shouted. Short of breath, his words were almost indecipherable, but the men halted their activity as they noticed him. He thought they seemed relieved.

Shyam's hands quivered as he held the pistol to Nizamu's head, a rare moment of hesitation for someone who usually favoured stoic execution. He looked at his company commander in disbelief. Had he asked him to stop?

No one moved. The pistol remained pointed at Nizamu's head.

'Stop!' Mohan yelled.

The company commander's instructions were clear now. Shyam, who had been seconds away from pulling the trigger, walked up to Mohan and hugged him. 'Sahib, thank you . . . you saved me from committing a crime today,' said the powerful hatchet man as he broke down.

'Fine. Get on with your job now . . .' Mohan wasn't finished, his gaze shifted to Nizamu. 'You . . . get up and go away.'

Nizamu appeared uncertain about Mohan's intentions, and wondered if he planned on killing him another way.

'Just run and don't be in my company area . . . I can't say what happens to you outside my jurisdiction. Run, and don't be seen again here.'

'Are you sure, haan (yes)?' Nizamu sounded doubtful. Was he, a man who was expecting death an instant ago, being given a chance at life?

'Jaao (Go)!' Mohan waved him away, signalling the encounter had ended. He had let Nizamu go. The former militant began to walk away in disbelief, occasionally turning around to look at Mohan and his men to confirm that he had indeed been granted a reprieve.

'You go . . . dikhna nahin . . . else goli maar doonga . . . (Don't be seen again. Will shoot you,' Mohan shouted at him.

Nizamu began walking faster, and then broke into a run. Mohan and his men watched him disappear in the morning mist.

Mohan turned to his men and asked, 'So what happened to Nizamu?'

'We shot him . . . sir.'

'And what happened after that?'
'We threw his body into the Jhelum.'
'Okay ... and this information stays between us.'

Life after militancy

Over the next week, Mohan kept an ear out for information about Nizamu. Hearing of an encounter in the neighbouring areas, he would call the concerned officers to check the names of the men who had been shot or captured. Nizamu never figured.

Three months passed, and Mohan prepared to return home on his annual leave. From Srinagar, he left for Jammu to board the Pooja SF Express train for New Delhi.

One of the most reliable ways to reach out to armed personnel in the area was via a phone call to the company commander and most villagers had the number. Mohan was used to receiving calls on this official line from sources and others who wanted help. Militants used it too. Mohan had once received a call from one who boasted about evading capture. The militant claimed he had been hiding behind a tree while soldiers searched for him.

At the railway station in Jammu, Mohan would hand the phone over to the JCO who accompanied him. The JCO would pass it on to the officer who replaced Mohan during his absence.

Mohan heard his phone beep and saw it was a call from an unknown number. He picked up the phone.

'Hello sir, do you recognize the voice? It's me, sir ...'

A subtle smile crossed Mohan's face as he enquired, 'How are you?'

'Fine. You're headed to New Delhi, sir?'

'Don't call me on this number. Here's my personal phone number. Call me on this,' Mohan directed.

Ten minutes after the train pulled out of the station, Mohan's personal phone buzzed. 'Hello sir, thanks for giving me your number. Where will you stay in New Delhi?' Nizamu asked excitedly.

'How do you know I am on leave?'

'Sir, I have been in this business for long . . .'

'Where are you now?' Mohan asked. The phone disconnected.

The train reached Paharganj station in New Delhi by 0500 hours. As Mohan left the railway station, he saw a shadowy figure in the morning mist, standing at a distance. The beard was gone, and it was a clean-shaven Nizamu that stood waiting for him.

'Do you realize the risk you're putting both of us in by being here?' Mohan asked.

'Don't worry, sir. No one will know. Look, I've been here for the last three to four months,' Nizamu said reassuringly.

He stood next to a sedan. Mohan said, 'Is that yours? When did you get a swanky car?'

'I was handling finances if you recall . . . I had told you about the pen drive, sir? The money I had hidden away from the militant group?'

'Hmm . . . so you got all the money with you?'

'I did, sir.'

Had he done the right thing in letting Nizamu go, Mohan wondered. He had asked himself this question several times over, and the thought now resurfaced. Each and every time, he came up with the same answer.

He had done the right thing . . . and his company men approved too.

'I've come to take you home.' Nizamu interrupted the silence, and his face lit up with a wide grin, the kind Mohan hadn't seen earlier.

The connecting train to Mohan's home town was to depart late evening, so the two men went to Nizamu's place. Several thoughts raced through Mohan's mind as he stood outside the house in a middle-class colony in New Delhi. Just six months earlier, he had stood outside another door, unsure of what lay ahead.

The door opened and Mehrunissa greeted them. It felt like a strange déjà vu.

Mehrunissa had prepared goshtaba, a Kashmiri delicacy made with tender meatballs in a fragrant yogurt-based gravy, for lunch. As they ate, Nizamu told Mohan more about how he had planned his escape from the militant group after meeting Mehrunissa. In charge of funds and accounts, he had moved money away in batches over months, risking being caught. Nizamu saw appealing to Mohan, a just and fair military officer respected by militants and villagers alike, as his way out.

By 1500 hours, it was time to leave for the railway station again. Nizamu offered to drop him, but Mohan insisted on taking an autorickshaw. 'I have a gift for you,' Nizamu announced and to Mohan's surprise, he brought a box with Old Monk rum printed on the outside. Inside were stacks of rupee notes.

Nizamu held the box out. 'This is for you . . . for saving my life,' he told Mohan.

'Is this how much you value your life? Is this what you think you are worth? Do you want to show me you have earned a lot? Or do you see it as a cheap transaction?' Mohan said, his tone laced with both anger and disappointment.

'Uh . . . oh . . . I am sorry . . . I didn't know how to repay your debt . . . I have no idea . . . you have no idea what you have done for me, sahib . . .' Embarrassed by his gesture, Nizamu folded his hands in an apology. The box was put back where it came from.

Mehrunissa then handed a basket of fruit to Mohan, who took it. 'Thanks . . . I love fruits . . . see, Nizamu, your wife knows how to gift . . . that's a schoolteacher for you,' he said and, to ease the awkwardness, patted Nizamu on the back.

That evening, on the train, Mohan thought about how much had changed in the last year. Would he be able to tell this story, ever? Maybe, some day. He was comforted by the fact that in Mehrunissa, Nizamu had found peace. When the train stopped at the next station, a young boy came to the window, selling tea. 'Chai, Sahib, chai?' he asked. 'Ek. Give me one,' Mohan said, putting all these heavy thoughts far from his mind.

For over two years since that meeting, Mohan resisted the temptation to find out what happened next in Nizamu's story. Occasionally, while passing through New Delhi, he thought of going to his house, but never did.

In due course, Mohan put in his papers for release from the army. He took up a job in a corporate firm in Bengaluru and went to work in the infotech hub in the city.

One day, getting into the elevator at work with all the giddy techies eager to kick-start a new week, Mohan received a text message on his phone. 'Moving to Canada. Wife is there. – Nizamu.'

In his office on the 10th floor, Mohan sat down at his desk and let out a deep sigh. Just as he was replying to the message, he received another. It read, 'I now have a son too. In Vancouver.'

Mohan walked to the coffee machine and poured himself a cup, gazing out of the window. The street below was bustling with cars, buses and techies hurrying to their workplaces in the infotech hub. Among them were ambitious young IT engineers dreaming of opportunities abroad. 'Some of them might be headed to Canada, not unlike Nizamu,' wondered Mohan.

Postscript

In the 1990s and early 2000s, scores of misguided youth would join the militant movement in Kashmir. The Indian government introduced a surrender policy in 1995 and then in subsequent years. But they didn't quite succeed. According to an Observer Research Foundation (ORF) report, political intervention, inadequate efforts in delivering promised incentives, social integration issues resulted in rehabilitated militants losing their way.

However, it's the young Indian Army officers on the ground, with their fingers on the pulse of militancy, who have played a greater role in winning back the trust of the youth in Kashmir.

10

Nariman House, 26/11: A War Comes Home

Of Bravehearts and their Families

On the evening of 26 November 2008, Lieutenant Colonel Sundeep 'Sandy' Sen and his wife Namrata had been preparing for their son's birthday celebration, scheduled for the next day. Just as they had finished dinner at home in the residential quarters of the National Security Guard (NSG) headquarters in Manesar, Haryana, Sandy received an urgent phone call. Colonel Sunil Sheoran, group commander of the 51 Special Action Group (SAG) and his CO, was on the line. 'Are you watching TV?' he asked.[1]

Sheoran was referring to the news of firing incidents in Mumbai, reported as a potential gang war in the city. On the blurry CCTV grabs, he had spotted spent cartridges on the floor of a café on Colaba Causeway and had known in his bones that this was no gang war. Gangsters don't use AK-47s, he thought.[2] When reports of a shooting inside the Taj

Mahal Palace and Tower Hotel came to light soon after, he immediately reached out to Sandy.

As Sheoran's second-in-command at the NSG – a counter-terrorism unit under the Ministry of Home Affairs comprising various units, including 51 SAG – Sandy enjoyed a warm camaraderie with his CO. The latter sounded genuinely concerned yet remained uncertain about the appropriate course of action for dealing with the Mumbai situation, since the intel coming in was sketchy. 'Why don't we do the drill and be ready?' Sandy suggested.

As Namrata sat down to help their son with his homework, Sandy went to his cupboard and pulled out his uniform. 'I'm going to the training area,' he said casually.

Living among the close-knit community of dedicated NSG personnel and their families in the residential quarters of the NSG headquarters in Manesar, Namrata was used to a life revolving around the training schedules and operations of the 51 SAG. A specialized unit designed to neutralize terrorist threats in vital installations, the men of 51 SAG excelled at handling hijack situations and rescuing hostages during kidnappings. Drills played an indispensable role in their training, instilling in the men a heightened state of readiness, with their responses finely tuned to expect the unexpected. So much so that, in the past, during an evening of drinks and dinner followed by a round of leisurely chatter, Sandy would playfully nudge the CO, subtly reminding him about the hooter, sounded to activate drills, all with a mischievous twinkle in his eye.

The night of 26 November, when the shrill hooter erupted like a fire alarm across the campus, everyone knew it was no joke.

Within fifteen minutes, men in backpacks and dungarees with rifles slung around their shoulders assembled in impeccable rows, ready to run 10 kilometres. Performing the drills, they prepared for the real operation that loomed ahead, in response to an unfolding situation that would eventually escalate into a national emergency.

Shortly after the initial incident, new reports emerged, revealing that it was not gangs but terrorists who had launched attacks at multiple locations in South Mumbai. Soon after, the entire news media went into a frenzy.

That evening at 2115 hours, ten armed men with rucksacks had stepped off an inflatable boat and walked into the city. They trooped off in groups to designated locations. Two of them had used maps to find their way past the congested Colaba market to reach a cream-coloured building. They waited anxiously. When the building guard took a break, they scampered in, unnoticed. Meanwhile, their compatriots had unleashed havoc in the city.

Visitors and commuters at iconic and bustling destinations like Leopold Café and Victoria Terminus (VT, now Chhatrapati Shivaji Maharaj Terminus) found themselves under a hail of bullets. Casualties had been mounting rapidly when word arrived that Hemant Karkare, chief of the Anti-Terrorist Squad (ATS) in Maharashtra, had been caught in an ambush and killed with senior police officials Ashok Kamte and Vijay Salaskar outside Cama Hospital, near VT.[3]

Rumours quickly intensified, spreading thick and fast, suggesting that terrorists were freely roaming the streets of Mumbai while the police seemed to be on the back foot.

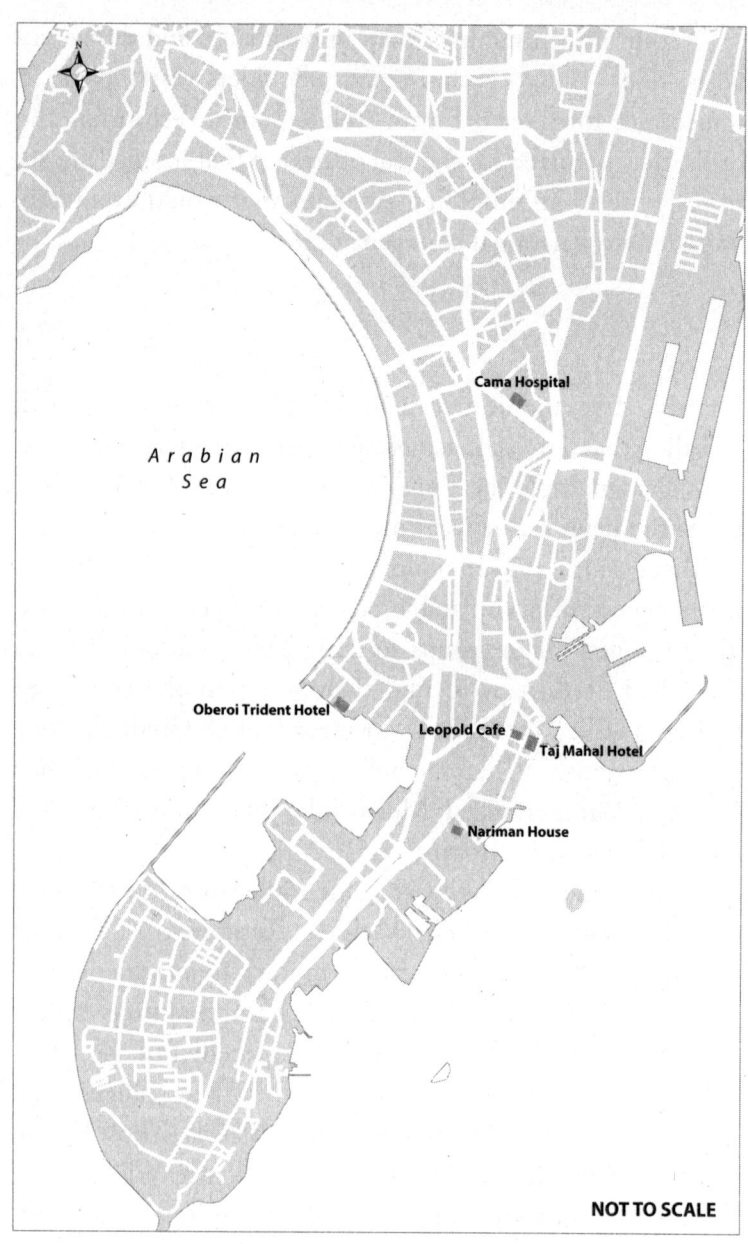

Some Locations of Attacks on 26/11 in South Mumbai

It had taken ten terrorists a mere sixty minutes to devastate the city[4] (see map on page 258).

27 Nov: Black Cats' Mumbai mission in Manesar

Back in Manesar, news of terrorists laying siege to the Taj and Oberoi hotels had been confirmed to the NSG team on standby. At 0018 hours on 27 November, NSG headquarters sent a code: 'Cheetah. Cheetah. Cheetah.' The NSG had been officially called in.[5]

Colonel B.S. Rathee, deputy force commander of the NSG, Colonel Sheoran along with an enhanced Counter Terrorism Task Force (CTTF), including CTTF-1 and a large part of CTTF-2, were flown in to Mumbai to be swiftly deployed.

Twelve hours after the terrorists had besieged Mumbai,[6] news of another hostage situation developing at Nariman House reached the authorities. Another NSG team was prepared for immediate deployment and readied to fly in. With Rathee already in Mumbai, Sandy was summoned to lead the makeshift CTTF-2 team.

As the sun beat down on the tarmac at the Indira Gandhi International (IGI) Airport, New Delhi, Sandy assessed the team of 148 personnel he was to lead before him, which was effectively only a reserve unit. It was an unusual group for this kind of critical operation because most of them were still undergoing training and were a month away from receiving the Balidaan badge.[7] The team was made up of men from the canteen stores department and storekeepers responsible for the army's retail stores and supplies. Additionally, sentries who were typically assigned to the quarter guard were present. Besides, the team included boys from the army mess, as well as helpers and those assigned to air marshalling and other non-combatant tasks. A ragtag bunch of fighters, Sandy thought, 'but they don't lack heart'.

Most of the key weapons had been sent with the earlier teams, so Sandy's men had to make do with what was left behind. There was a shortage of MP5 submachine guns, so pistols assigned for air marshal duties were carried. At the airport briefing, Sandy assigned authority roles to individuals within the team, appointing 'hit commanders' and 'leads', mindful that the team had very little experience in hardcore operations and never actually practised and operated together as a team before, which is a critical aspect of counterterrorism operations.

Right off the bat, Sandy was confronted with his first challenge of leading a hastily formed team. The IL-76 – an airship designed to transport large groups of soldiers along with weapons and equipment – had stood for an hour awaiting the arrival of Major General Abhay Gupta, inspector general (IG) in charge of NSG operations, who was flying in from Mussoorie by helicopter and unable to land because of ground haze.[8] Sandy also faced a shortage of one officer as Captain Mohit Dhingra had taken leave. He had gone to Dehradun with his fiancée Keshar to introduce her to his parents. When he received Sandy's urgent call about the operation, Dhingra left Dehradun immediately. However, en route to IGI Airport, his vehicle broke down so he had called Sandy to request additional time.

'We must leave now,' the IL-76 pilot told Sandy.

In the meanwhile, Mohit had managed to flag down a Maruti Gypsy SUV on the highway and convinced the driver to take him to the airport.

Sandy's men were already on the plane. After Gupta's arrival and boarding, Sandy gave the signal to the pilot that they were prepared to depart. His watch showed the time was 1000 hours. Too bad Mohit couldn't make it on time, Sandy thought. A diligent team player, he would be missed.

As the plane taxied on the runway, Sandy spotted a man in black dungarees, carrying a backpack, racing towards the departing plane. It was Mohit. Sandy ordered the plane to be stopped and the crew ladder to be lowered.[9] Once Mohit scrambled on board, the IL-76 aircraft soared into the blue skies, heading towards Mumbai.

As he watched his mates kit up during the flight, Sandy couldn't help but reflect on the hard work and persistence that had led up to this moment.

Two years ago, he arrived from a United Nations posting and proceeded directly to the Military Secretary's office in the Army Headquarters, New Delhi. After several successive field postings, he now sought a staff posting to enhance his career prospects. With determination, he presented his case, momentarily surprising the Lieutenant General at his desk.

A couple of months later, when Sandy's posting order arrived, it was a surprise. He was being sent to the NSG. The qualifications needed to become a commando were demanding, especially for a man in his late 30s, but Sandy pushed himself hard to meet the fitness requirements. At the time of the Mumbai operation, he had waited for a whole year to be involved in action. Little did he know that he was about to become part of the narrative of India's biggest terror strike.

As he gazed at the younger officers on the flight, Mohit and Major Manish Mehrotra, who had hurriedly returned from a short holiday in Gurgaon to rejoin his unit in Manesar, he was reminded of his own motivation for being on this mission. Both had made efforts to be part of the operation despite personal commitments. Their dedication and willingness to take on challenges for the mission at hand inspired Sandy to think, 'This bunch is special.'

Namrata and the families in Manesar

As Sandy's plane flew through the skies towards ground zero, his wife Namrata had woken up late that day. She was busy with household chores when Lance Naik Dinakaran, the sahayak (assistant), arrived to inform her that Sen saab was on his way to Mumbai. Namrata had been half-asleep when Sandy left the house early morning, and vaguely recalled him whispering in her ear that an attack had taken place. Years of living with an infantry officer had prepared her for unexpected moves, so she didn't think much about it. Helping her son with his studies, she hadn't watched the news and had no idea about the big attack on Mumbai or her husband's role in countering it.

That afternoon, she received a frantic call from Ekta Kandpal, the wife of Major Sanjay Kandpal. The Major was in charge of one of the 51 SAG squadrons, which was part of the first NSG team sent to Mumbai to counter the Taj hotel attackers.

Ekta had sobbed uncontrollably on the phone.

'What happened?' a puzzled Namrata asked.

Her voice trembling with fear and urgency, Ekta described

the magnitude of the attack in Mumbai. She repeatedly uttered, 'Sanjay is gone.'

'What?!' Namrata's initial response had been one of disbelief and shock. Soon, she recognized that Ekta was just unnerved about her husband being in the midst of an escalating crisis, so she quickly tried to calm her down.

The families of the NSG units started a vigil in front of their TV sets, absorbed in the bombastic minute-by-minute news coverage. Despite the intense reporting, there was no information about their husbands.

Worries grew deeper. That evening, Namrata and a few families connected over a conference call, trying to support each other and ease their collective fears. Deep down, Namrata was beginning to feel uneasy. She spent the evening surfing the internet, following updates and information about the battle that was taking place in Mumbai, for any word about Sandy.[10]

Fully engaged like everyone else, staying informed had become crucial to her during this crisis.

A house with no address

In 1993 and 2006, terrorists had carried out bomb blasts in Mumbai, leaving devastation in their wake. However, in those cases, the actual attacks had ended in a matter of minutes. This time was different. The ongoing assault was daring and clandestine, and the terrorists seemed to have the upper hand. Yet, a glimmer of hope emerged as NSG teams arrived at the Taj and Oberoi, raising people's collective spirits.

News channels continuously reported on the composition and expertise of the NSG. As a result, everyone in the city started googling this crack anti-terror outfit. It garnered more admiration than even cricketers and film stars. The public's hopes soared as the army remained the last bastion of the fightback.

Thirty kilometres away from the chaos of the attacks in South Mumbai, an army plane landed at Sahar airport at 1200 hours on 27 November. A dozen BEST buses and a car were ready on the tarmac to receive the NSG team.

Gupta told Sandy he was heading to the Taj, where NSG action against terrorists was under way. Sandy was sent to Mantralaya, the state secretariat. 'You will be briefed there,' Gupta informed him.

Over a hundred people welcomed the commandos at the airport. Sandy had never seen such a large reception before. As they headed into the city, the commandos waved at the cheering crowds who lined the road.

As his Tata Indica car took the Peddar Road flyover, Sandy looked at the skyscrapers that surrounded him. For long, he had wanted to visit the mighty financial capital and India's most international city, under different circumstances, of course.

After an hour-long ride, accompanied by a JCO and a Mumbai police inspector, Sandy arrived at his destination.

On any given Friday, Mantralaya, a seven-storeyed building, is packed with hundreds of people milling around on its premises. Bureaucrats, politicians, clerks, officials, press people, hangers-on – it's a stomping ground for power brokers in the state. On that day, as Sandy approached the entrance, there was a pervading eerie silence. He stepped inside the building and found the corridors cold and empty.

'Koi hai (Is anyone there)?' he asked and heard his voice echo in the dark corridor.

On further investigation, Sandy discovered a lone bulb illuminating a corner room on one of the floors. Inside, a scrawny figure sat hunched over a small table.

'Saab, yahan toh koi nahin hai (Sir, there is no one here),' Sandy was told.

Mantralaya had been deserted. The power brokers were gone.

Sandy called Gupta, who was at the Taj hotel and had some crucial details about the location. 'Sundeep, it's an old age home. And I think two militants have gone there . . . they have taken some hostages. Maybe a few elderly guys, you know . . .'

Sandy pictured an old age home as a serene villa with a beautiful garden surrounding it, much like the one depicted in the Hindi film *Munnabhai MBBS* that he had seen a couple of months ago. He wondered if the target was challenging enough and was confident that he had enough men to handle the task.

Before ending the call, Gupta repeated the name of the target location, which was Nariman House.

'Sir . . . no one is here to tell us where that is,' said Sandy, as he and his men were the only people on the Mantralaya premises. 'Check with someone at the Colaba Police Station. They will guide you,' Gupta assured.

Sheoran, who had arrived before Sandy, had deployed his HIT (house intervention) teams at the Taj. Operations were under way and a group of four teams had been put under the charge of Major Sandeep 'Unni' Unnikrishnan to storm the older of the two Taj buildings.[11] Before heading out to

Colaba, Sandy received a call from Sheoran, who urgently needed reinforcements to storm the Taj hotel. Fifty men were dispatched to support the three officers and the over a hundred NSG men already deployed there. Sandy had a makeshift reserve force to begin with, and now he was also short on numbers and ready to mount an operation in a location no one seemed to know!

On the way to the Colaba Police Station, Sandy noticed a young, lean man in a red T-shirt and shorts driving a scooter alongside their contingent. He had seen the scooter earlier but didn't give it any consideration. They entered the police station and found it deserted, just as Mantralaya had been. It was like a scene from the Hollywood movie *Apocalypse Now*!

Here, too, they came upon a lone havildar in the building, who didn't know where Nariman House was. A normally unruffled Sandy found his patience wearing thin. As he stepped out of the police station, he saw the boy in a red T-shirt waiting at the gate. He appeared keen to speak but was scared and hesitant.

'Hey! You come here' Sandy called out. Getting straight to the point, he asked, 'Do you know where Nariman House is?'

The boy nodded nervously and stammered, 'Sir, that is what I wanted to tell you for a long time! I know it, sir . . . I will take you to Nariman House.'

Taken aback, Sandy had a question first. 'And who are you?'

Spectators at the scene of combat

The boy identified himself as a local Jewish resident, representing one of the few remaining members of the community in Mumbai. India has a minuscule number of Jews, most of whom reside in the city.[12] He informed them that Nariman House was close to the Colaba Police Station, located on Hormusji Street, just off the Colaba Causeway market area.

Hormusji Street is a narrow lane, with a cramped cluster of closely packed buildings making it difficult to tell one roof from another.

When Sandy and his SAG team reached the location at 1300 hours, they were greeted by a sight that shocked them. Although the Causeway and shops had been closed down following the terrorist attack, a large crowd of a few hundred people had gathered on the roads and buildings surrounding Nariman House. A far sight from the ghostly corridors of Mantralaya and the Colaba Police Station, there were people everywhere – on the streets, atop the shops, craning through windows, perched on buildings that line up the street.

Amid reports of an ongoing terrorist attack in the city, the initial sound of gunfire was what drew people to the narrow Colaba by-lane on the night of November 26. Smoke billowed from the window of a building known as Nariman House, where a terrorist emerged, opening fire and even throwing a grenade on to the street, which further attracted onlookers. Simultaneously, a bomb planted by the terrorists at a nearby petrol pump detonated. Subsequently, a fearful group of locals and police had gathered in a chaotic vigil, awaiting reinforcements. Much to their chagrin, the NSG commandos were welcomed with cheers, shouts and whistles. Entering the area, Sandy felt like the legendary cricketer Sachin Tendulkar walking in to bat at Mumbai's Wankhede Stadium, as if a thousand eyes were on him. The capacity crowd surrounding the building with terrorists, whose aim was to inflict damage, made an already dangerous situation perilous. Moreover, no one had any idea of how many terrorists were in the building or the extent of weaponry they were armed with.

Several high-ranking Mumbai police officers sat by the street next to the building, waiting for their luck to change. Sandy approached them and said, 'Tell me what the situation is.'

The officers appeared bewildered, unable to fathom how the situation was developing. Sandy had more questions. 'Where is Nariman House?' 'How many people are staying there?' 'Where are the terrorists?'

Now the responses trickled in but were inconsistent. A few officers claimed there were four militants in the building. Chaar (four), Sandy repeated, gesturing with his hands. 'Yes,' he was told, even as the policemen looked at one another uncertainly.

'So, how many, sir?' asked a news reporter who stood nearby.

'If you let us do the work, you will know at the end,' Sandy curtly replied.

The reporter turned to the camera and delivered his verdict. 'No one knows how many militants are in there.'

Huddled together, Sandy and his team created a rough sketch of the area around the building. In contrast to the terrorists, who were subsequently discovered to possess detailed maps of the locality, the NSG had to draw their own. They then formed a cordon, to ensure that the terrorists

Sandy (centre), Major General Gupta and others discuss the plan outside Nariman House

wouldn't slip away in the commotion. A sniper detachment was positioned at a vantage point across Nariman House, ready to pick out terrorists if they attempted to sneak out.

The commandos could not see inside the house, though. In fact, no one in the vicinity knew much about the layout of the house. They were told that earlier in the day, a small boy named Moshe, along with his nanny, had fled the house to safety and gone to the police headquarters. Moshe Holztberg is the two-year-old son of Rabbi Gabriel Holtzberg and Rivka Holztberg, emissaries of the Chabad Movement – a Hasidic Jewish movement known for its efforts to promote Judaism – in Mumbai. However, at the time, since the cops did not question the nanny, information about Nariman House's occupants or structural design remained unclear.

'How many people are inside?' Mohit asked. Bystanders

offered opinions. 'Ismein militants pehle se rehte thay (militants were living here earlier),' 'Rehte militants hain, aapas mein jhagda ho gaya (the militants who live here are fighting amongst themselves).' However, this was just speculation. The boy in the red T-shirt informed Sandy that it was a Jewish household, and the locals had mistaken the Jewish men, with their beards, for Islamist militants.

The boy guided Sandy and his men to the building, which turned out to be nothing like what Sandy had imagined from *Munnabhai MBBS*.

On the third floor of Merchant House, a building within arm's reach of Nariman House, lived Sandeep Bharadwaj, CEO of Tower Capital and Securities Private Ltd, a private investment

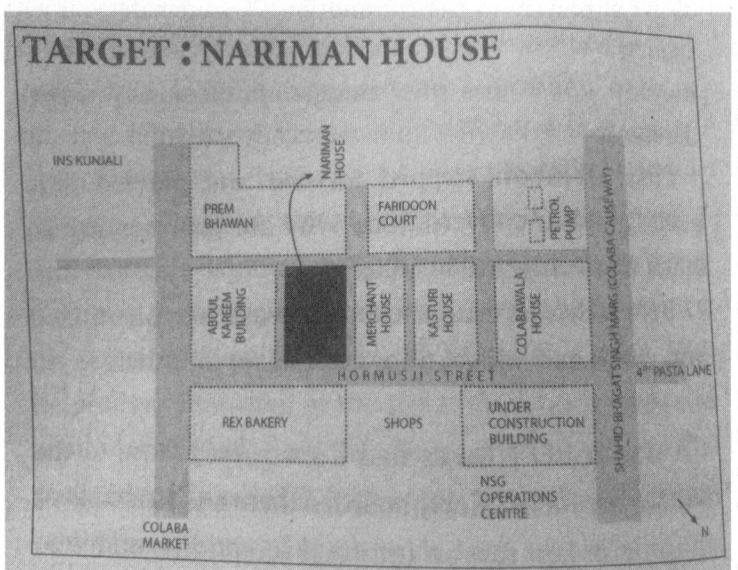

Nariman House and the neighbourhood

banking firm, with his British wife Lucy. The Bharadwajs had sought refuge in a corner of their house since the previous night, having survived a barrage of bullets and the impact of the bomb planted by the terrorists.

As they tried to remain quiet and inconspicuous, Lucy's phone suddenly rang. It was the SAG team, attempting to contact individuals for rescue and neighbourhood evacuation. The local municipal corporator, who had the couple's phone numbers, had provided them to the police at the scene, who, in turn, had passed them on to the SAG team.

Lucy, who had been a reservist with the British Territorial Army while in college, would prove to be an important ally during this time. Her training had helped her stay unruffled and make crucial observations during the ongoing events. Despite taking cover, she carefully watched the firing coming from the window of Nariman House. Both Sandy and Mohit spoke to her. She informed Sandy on the phone that the terrorists had hurled 15–16 grenades and provided an approximate count of the ammunition they had fired until that point. She was almost certain about the number of terrorists in the building. 'Two,' she said, having witnessed them firing from the window. The terrorists had attempted to deceive the police by firing from different points, but a vigilant reservist saw through their ruse.

Sandy now had a clear answer about the number of terrorists his team would have to tackle.

Changing tactics: From operation rescue to seek and destroy

As the last light ebbed away and darkness fell, the commandos planned their move. Electricity to Nariman House was cut off.

At around 2100 hours, a five-man team tried to blast their way into the building. But the attempt was short-lived. The terrorists responded by lobbing a grenade in their path and intermittently firing away at the advancing team. As a frontal attack was thwarted, the commandos realized the only way into the building was from the terrace.

The plan was to storm the building and launch a hostage rescue assault, scheduled to commence at daybreak. This timing would minimize the risk of shooting hostages, as snipers would have a clearer view of the inside of the building. In the meantime, the commandos would carefully sweep the buildings surrounding the area, aiming to avoid any civilian casualties.

Half an hour after the lights went off, the phone rang.

In Nariman House, the mood was grim. Lucy's hunch had been right – there were two terrorists in the building. It was later learned that Babar Imran, alias Abu Akasha, and Nasir, alias Abu Umar, had been acting on orders from Sajid Mir, their handler in Pakistan.

'Get rid of them (hostages). Firing could start on you at any time and you risk leaving them behind,' Sajid had warned Babar.

An hour later, Sajid called to check if the job was done. 'I was waiting for your call . . . I will do it,' Babar pleaded. A single shot was fired.

'Was that one of them (hostages)?' asked Sajid. 'No . . . both together,' Babar fumbled his reply. 'Inshallah,' Sajid said and signed off.

This conversation was intercepted by Indian intelligence; however, inexplicably, the information was not passed on to Sandy and his SAG team at Nariman House.

By an hour past midnight, Mohit, Manish and their teams had completed the evacuation of a dozen families, including the Bharadwajs, and around sixty people from the vicinity.

Sandy established a command base on the roof of Prem Bhavan, a six-storey building situated at the rear of Nariman House, providing an excellent vantage point and a clear view of the target location.

As he looked at Nariman House in the darkness, a strange voice caught his attention. Was it from Nariman House, he wondered. Initially, he didn't quite think so. But then, he cocked his ears to figure out what was happening.

A muffled voice of anguish reached his ears. It was a woman's voice from inside Nariman House. Sundeep couldn't figure out which floor it was coming from. In the cool air, the voice seemed to rise and fall, echoing intermittently from the building. 'Who could that be?' wondered Sundeep.

At 0500 hours, Sen briefed his twenty-man team, dividing them into four squads, with Mohit and Manish leading two squads each. At dawn, they would storm Nariman House from the rooftop.

Just before the operation commenced, Sandy received a call from Gupta. The police, relying on Intelligence Bureau sources, had received a critical tip that would change the course of the operation. They were informed that two militants were inside the building and had already killed their hostages.

The operation's objective had shifted from rescuing hostages to seeking and destroying the targets.[13]

Hunting the target, a man down

At the crack of dawn on 28 November, the first helicopter with Mohit and his team took off from INS Kunjali (now INS Shikra), the naval air station in nearby Navy Nagar, and arrived over the building.

Due to the strong winds, the pilot struggled to stabilize the helicopter, and its rotor nearly struck the building's corner as the aircraft lurched and swayed closer to the roof. The gunner, who sat alongside the pilot, threw down the rope for the commandos to slither down. Mohit had practised the drill innumerable times and he descended with assurance and speed. Except, he had climbed down to see Sandy standing in wait. He had lowered himself on the wrong roof! Damn! Without available maps with closely packed buildings spaced around 10–12 feet apart, the gunner's rope, swayed by the winds, landed on the adjacent roof.

'Itna drama karke upar se kyun aaya? Seedhe mere saath stairs se yahaan chad jaata (What was the need for this dramatic entrance? You could have just taken the stairs with me),' said Sandy, shaking his head in mock admonishment. This light-hearted moment eased some tension. 'Now, go back, take another helicopter and return. This time, land on bloody Nariman House!' Sandy ordered.

Mohit and the team ran down the stairs and hailed a couple of black and yellow taxis to rush to INS Kunjali. Meanwhile, a second helicopter hovered over Nariman House, allowing Manish and his men to descend on the correct roof.

All this time, TV channels were busy telecasting the helicopter operation on their channels. An enthusiastic crowd cheered on from the ground, watching the proceedings. No

Nariman House, 26/11: A War Comes Home

Commandos being helidropped on the roof of Nariman House

one realized that the first landing had been on the wrong roof, not even the terrorists.

Like everyone else, Sajid Mir was watching the event live on his TV. He called Babar's phone.

'Heli aa gaya upar (Can you hear the helicopter above you)?' Sajid asked tensely.

Babar confirmed. 'Humare upar fire shuru ho gaya hai. Shuru ho gaya (They have started firing at us, they have started),' he kept repeating.

'Take cover,' Sajid instructed them.[14]

The Pakistani handler was providing specific details to the

terrorists about the landing of the commandos on the roof. 'Pandrah bandey aa gaye chhat par (Fifteen men are on the roof)... stay in the room but keep the stairs in your crosshairs, targeting the commandos as they come down from the roof,' Sajid instructed Babar.[15]

The live TV telecast had revealed the number of commandos that were part of the operation and the weapons they carried. Sajid instructed Babar and Nasir to take 'cross positions' in the 300 square foot living room, covering the entry points on opposite sides. Using sofas and the refrigerator, they created barricades at both ends of the room and positioned themselves behind them. Only their rifle barrels poked out of these makeshift bunkers, providing cover for the door from both sides. The single-entry point to the fourth floor turned into a hazardous kill zone.

The commandos descended to the fourth-floor landing of Nariman House. The door to the floor was locked, but an explosive charge blew it open, allowing the commandos to barge in.

Havildar Gajender Singh Bisht charged in with his MP5 submachine gun. Nasir, crouched behind the refrigerator, opened fire with his AK-47. Bisht was hit; falling to the ground, he continued firing.[16] Manish and the team retreated, ascending the stairs as a melee of gunfire rocked the building.

Sandy had heard both the MP5 and AK-47 guns go off and knew his men had located the terrorists. Just then, he received a message that Bisht was down. He was concerned that the terrorists would take Bisht hostage or, even worse,

Nariman House, 26/11: A War Comes Home

NSG Commandos prepare to attack from the rooftop of Nariman House

string his body up from the wall of the building for the TV cameras.

'They must not get him,' Sandy yelled to his men.

Meanwhile, Babar had called Sajid Mir in Karachi. 'I've been shot . . . I've been shot,' he moaned.

'Kahaan lagaa (Where were you hit)?' Sajid asked.

'One on the arm and one on the leg,' said Babar.

'Did you kill anyone?' the handler asked. Babar confirmed he had killed an NSG commando.

'Alhamdulillah (Praise be to God). May God keep you with him,' Sajid said.[17]

To maintain the pressure on the terrorists, Sandy used a tear gas gun to fire shells into the fourth floor of Nariman House. A wisp of gas emerged from it. He fired again, and the room filled with gas. A choking Babar wetted a pillow and struggled to get to the window. Aware of the danger, he aimed his AK-47 at it.

Every move was being guided by the handler, sitting miles away. 'Can you see any faujis (Indian soldiers) outside?' asked Sajid. 'Saamne sniper nazar aa raha hai (I see them) at the window,' Babar replied. 'Maro . . . Maro fire unhe (Fire at them!)' instructed the handler.[18]

A few moments passed; Sandy wasn't able to see much as gas continued to billow from the window. Looking intently, he could faintly see Babar's eyes at the window.[19]

As Sandy craned his neck to get a closer look, Babar aimed his AK-47 and fired at Sundeep. *Crack!!*

Meanwhile, back in Manesar

Army wives have coped with their husbands' postings in difficult areas before. But their husbands' dangerous mission being shown live in their living rooms was a new experience for them.

The images were devastating – the burning of the Taj hotel by terrorists, killing of police officers and newscasters chattering on. The uncertainty faced by those inside and the powerlessness of those outside. Terrorists had seized control of locations in Mumbai, while in Manesar, families were held captive by TV news.

The ladies of Manesar shared an unspoken worry: Whose husbands might have been at risk of losing their lives? Whom might they have had to console soon? The fear was real.

It's almost as if they had known devastating news was an hour away.

Soon after the NSG commenced their first attack on Nariman House, on the morning of 28 November, Namrata received another frantic call from Ekta, who had been anxious since her husband left for the operation.

'Have you heard? There has been one death . . .' said Ekta, then broke down, sobbing.

Namrata baulked at the prospect but decided to cut short the conversation. Following the news ever since she had heard Sandy was in Mumbai, she did a quick search online.

Breaking news! flashed on the screen. Commando Sundeep nahin rahe. (*Commando Sundeep no more.*)

She carefully read each word of the Hindi bulletin in Devanagari script again; no, she hadn't read it wrong. The screen went dark before her and overwhelmed, she collapsed and tumbled from her chair.[20] Hearing the loud thud, Dinakaran rushed into the room and saw Namrata on the floor, possibly unconscious.

'Memsaab! Memsaab?' he shrieked.

On receiving no response, he came closer and checked her pulse, and then rushed to the phone. A soldier trained for emergencies, he remained composed and dialled the unit doctor. Namrata remained unresponsive as Dinakaran moved her to the couch. He then called a few of the officers' wives who had been in touch with Namrata over the last few days.

Ekta rushed over, and a few other ladies followed. The Sen household was filled with nervous conversations. The doctor hadn't arrived yet. Dinakaran called again and was informed that the doctor was on his way. In the meantime, Dinakaran made arrangements to ensure they were ready to move Namrata to the unit hospital, if required.

The unit doctor arrived and checked Namrata's vitals. 'She has fainted,' he told the circle of people gathered around her. A few moments later, Namrata regained consciousness. She was shaken up, and her thoughts were blurred. She heard one of the ladies say it was Major Sandeep Unnikrishnan who had died, and not her husband.

Despite her efforts to compose herself, Namrata remained unconvinced. They showed her the news on the phone. Sandeep Unnikrishnan – a young officer highly esteemed by her husband – had perished. Namrata's immediate reaction was one of quiet relief that it wasn't her husband. Then, reality set in, and she grieved for the young officer whom she had known well.

Sandeep Unnikrishnan was one of the best and he was dead. Namrata couldn't think clearly and was both relieved and sad. She was scared for her husband now and confused.

Grieving for fallen comrades even as battle continues

Back in Mumbai, Babar's gunfire at Sandy ricocheted off the window grille and set off sparks. The bullets missed him. Sandy took cover, got to his feet and fired another tear gas round. The shells knocked down the grilles.

'Laddoo Lao,' he yelled to his men.

He was handed a HE 36 hand grenade.[21] Sandy pulled out the pin and tossed it, and a shattering explosion ripped through the room. He wasn't done yet. He lobbed a few more, and blasts and smoke filled the air. Inside the building, Havildar Ram Niwas used the distraction to throw a fire hook on Bisht's lifeless body, and the commandos proceeded to carefully haul him up the stairs.

Nariman House, 26/11: A War Comes Home

Sandy firing at terrorists inside Nariman House

Meanwhile, Mohit received a phone call.[22] It was his fiancée, who sobbed as she informed him that Major Sandeep Unnikrishnan had died while rescuing hostages at the Taj hotel, caught in a crossfire with the terrorists.

Unni's team had been rushing up the stairs to the first floor next to the Palm Lounge when they were ambushed by the terrorists. The commandos found themselves pinned to the base of the staircase by heavy gunfire. However, Unni and fellow commando Sunil Jodha had paired up to fire and move up the stairs. A bullet struck Jodha on his abdomen and he fell. Unni dragged Jodha to the side, gave him cover and kept firing from his MP-5.[23] The brave officer then charged alone, going around the Palm Lounge. A terrorist hidden in the folds of the wall shot Unni.[24]

'Major Unni is dead!' Mohit, who knew Unnikrishnan well, was overcome with emotions and vowed to finish off the terrorists. Manish advised him not to be foolhardy. Mohit had a plan: as an engineering officer assigned to the NSG, he was a specialist in the use of plastic explosives.

He suggested they blast their way in. Sandy, who didn't want the terrorists to hold them off until sunset as the siege might then continue the next day, consented. The team quickly drew up a plan to launch the assault.[25]

Even as the battle was under way, Sandy phoned Namrata and requested her to go to Bisht's house and break the news of his death to his wife.

Namrata was still reeling from the false news about Sandy, and the reality of Major Sandeep Unnikrishnan's death, but she put up a brave front for the sake of Bisht's wife and children.

Breathing heavily and with her heart pounding in her ears, she arrived at the Bisht home.

Bisht's wife stared into Namrata's eyes as she attempted to soften the blow and gently convey a partial truth.

'You may have to go to Mumbai ... Gajender has been hit by a bullet and he will need some time ... err ... ' Namrata trailed off unconvincingly.

Bisht's young wife clutched her children, desperately wanting to believe Namrata but searching for clues in her words. Just then, a car pulled up outside. It was Mrs Gupta, the General's wife. Mrs Bisht looked at Namrata, and then at Mrs Gupta who had entered the house. 'Why is she here? Why are all of you here? I know why. I know ... ' she said, before breaking down hysterically.

No one said a word. Mrs Bisht had crumpled on the shoulders of one of the visiting women.

A few of the wives summoned the strength to speak with the spouses of the younger officers and jawans and offer words of advice and support.

Everyone had been aware that operations were under way. Nonetheless, there was a noticeable shift in emotions: fear had transformed into gloom following the two deaths.

Namrata's worry had become palpable now. Images from war films flashed through her mind. A soldier lighting up a cigarette and being shot in war. 'Don't call your husbands, please!' she told the women assembled, haunted by the image of Sandy's Android phone lighting up in the dark during a firefight – a potential giveaway to the terrorists.

The final attack

Meanwhile, unbeknownst to the media, government, police and the families, the final stage of action was being readied in Mumbai.

As Mohit created a plastic explosive out of a 2-kg slab of PEK,[26] Sandy climbed up a two-storey building about 100 metres from Nariman House and used an AGL to shatter the windows, letting the light in to improve visibility for the commandos within.[27]

An AGL is rarely used in an urban setting, and this was probably the first-ever use of an AGL in Mumbai. Two soldiers prepared the weapon, and Sandy gave the order to fire.

As the grenades struck the building, shattering the windowpane, Mohit's explosive blasted a three-foot-wide hole into the wall on the fourth floor. The commandos swiftly barged in.

They found Babar Imran slumped against the wall, his face

coated with plaster, dust and paint. He looked tired, but ready for a final hurrah. One of the SAG commandos, Lance Naik V. Satish, was 5 feet away. Babar fired and missed as Satish ducked and fired at the same time. He found his target; Babar fell backwards. Satish fired again, and Babar's AK-47 fell from his hands. He was killed. Nasir had died earlier and the commandos retrieved his burnt torso.

More commandos entered Nariman House. It was all over, finally. The bodies of six executed hostages were discovered. Some had been bound and blindfolded with strips of cloth, their faces and bodies mutilated. It was a gut-wrenching sight to behold.

Sandy's motley, ragtag bunch had emerged victorious against the monsters, and they could now take solace in their hard-fought triumph.

The missing terrorist

An exhausted Sandy returned to his hotel to catch some sleep. He called Namrata, and said, 'I am safe and sound, mission completed.'

Namrata had been waiting for news, surrounded by the families of jawans, including Havildar Gajender Singh Bisht's wife. Hearing Sandy's voice, she heaved a sigh of relief.

After speaking to her, Sandy crashed on his hotel bed. The phone rang at 0230 hours and it was Sheoran. Sandy was needed at the Taj hotel. He got up and dragged himself to the car in the foyer.

The final battle was under way when Sandy reached the hotel. The guests had been evacuated and the terrorists were being taken out. Half an hour later, Sheoran informed Sandy that all terrorists in the Taj attack were dead, barring one, who was missing. 'Has he escaped into the city? . . .what if . . .?' a worrying thought crossed Sandy's mind. He, along with other commandos, began a search of the premises.

The hotel had been set on fire earlier during the attack, and there were firefighters and commandos everywhere – walking under collapsing beams, broken furniture, ruptured ceilings and burnt interiors.

In the Harbour Bar, located in the north wing of the hotel, firefighters had been clearing the rubble when Sandy noticed a rifle barrel sticking out from the debris. He pushed it aside to discover a tiny fragment of a body buried in the rubble. 'Two kilos of flesh – that's how a monster is remembered at the end of a mission,' Sandy thought.

The count was complete. The terrorists were all dead. Both hotels had been cleared. On 29 November, the city had been won back.

Outside the hotel, Sandy stared at a signboard – Ramchandani Road – named after Flying Officer P.R. Ramchandani, a gallant air force officer who died in the war against Pakistan in 1965. Many years later, Major Sandeep Unnikrishnan laid down his life fighting terrorists from Pakistan on Ramchandani Road. Two individuals, separated by time and different conflicts, but linked by their sacrifices on the same road. A poignant coincidence, Sandy thought as he, Sheoran and others sat alongside each other in silence.

A startling revelation

The next day, the NSG teams were back in New Delhi. India's biggest urban counterterror operation had just ended successfully.

A few days ago, when the first team was departing for the mission in Mumbai, Home Minister Shivraj Patil's delay had held up their flight.[28] On their return from Mumbai, Sandy and the others didn't hear from any head of the ministry, government or the Indian Army. It was a quiet return to Manesar for the warriors.

Five days later, Sandy was at home in the evening, watching TV with a glass of whisky in hand. The channel was playing a video tape from the attacks. They showed a conversation in which Sajid Mir, the handler in Karachi, was imploring Babar to kill the two hostages in Nariman House. However, there was only one shot heard. The information, relayed by Indian Intelligence to Mumbai Police and then to the NSG earlier, was that both hostages had been killed.

Sandy glanced at the recorded time of the conversation displayed at the corner of the TV screen. There was a lump in his throat. The call was made a few minutes before he heard the muffled cries of a hostage on the terrace of his command post next to Nariman House. He had not been aware of this particular conversation intercept between the two terrorists. All along, the SAG unit had trained to save a hostage over killing a terrorist. He and the commandos didn't carry out the intervention thinking all hostages were dead. Did the terrorist lie to his handler? Was the hostage alive when he heard the cries? Why did no one bother to tell the NSG about the conversation intercept?

Sandy took off his rimless glasses and shut his eyes for a moment. He could hear the house gate opening, and his car entering the porch. Namrata had returned from another visit to Havildar Gajender Singh Bisht's home.

Postscript

This postscript includes information that emerged after Operation Black Tornado, the code name given to the NSG's counterterrorism operation against the deadliest terrorist attack on Indian soil. It aims to piece together the backstory, providing readers with a more comprehensive understanding of the events that unfolded in this chapter.

In 2006, a man named Daud Gilani arrived in Mumbai carrying his mother-in-law's camera, two credit cards and US$ 4,000. He returned in 2007, then April and July 2008 as well. Going by the name David Headley, he was Mumbai's quintessential invisible tourist – a white man with camera in hand marvelling at the sights. In reality, he roamed the city and took pictures and videos that were handed over to his ISI handler Major Iqbal and Sajid Mir from Karachi.

Iqbal and Sajid used Headley's information to sketch a map of South Mumbai.

On 18 September 2008, a satellite conversation between a Lashkar-e-Taiba (LeT) operative and an unknown caller was picked up by India's Research and Analysis Wing (RAW).[29] There was a plan to target a hotel at Mumbai's Gateway of India via the sea. Six days later, on 24 September, RAW picked up another chat. The LeT man mentioned the targets. The Taj hotel, Marriott and Taj Land's End figured in the chat. On 19 November, the plans were more specific over another phone call. RAW intercepted this one too. A chilling message stood out on the call: 'We will reach Bombay between nine and eleven.'

Clues kept popping up. The Intelligence Bureau received news of a suspicious boat on the Arabian Sea. It marked the file 'top secret and confidential' and passed the message to the Coast Guard and Indian Navy headquarters. The Coast Guard launched a search for the Pakistani boat but didn't find one. The Indian Navy was indifferent since the boat was in Pakistani waters. A Navy Admiral scrawled 'No further action' on the message. Evidently, the second half of the message which was ignored read: ' . . . the boat is attempting to make an infiltration.' Ironically, ignoring the vital part of the sentence made the information appear banal.[30]

Six days later, on 25 November, 'the boat' infiltrated Indian waters. The occupants – ten armed terrorists steered over wireless radio sets by their handler in Karachi, Pakistan – planned to be on Indian shores, dedicated to completing a deadly mission.

Two of the men had been assigned the only non-commercial address on the list. It was neither a hotel, restaurant nor railway station. It was also the only address among the targets that was unknown to locals. It was ironic that while the police and intelligence agencies were unaware of Chabad House, Headley possessed knowledge about it and specifically selected it as a target for the attack on the Jewish community.

Nasir alias Abu Umar and Babar Imran alias Abu Akasha entered Nariman House (the name of the building designated a Chabad House or Jewish community centre) and proceeded to kill three Jewish house guests – Yocheved Orpaz, 60, Bentzion Chroman, 28, and Rabbi Leibish Teitelbaum, 37. They took four hostages: the Holtzbergs, their son Moshe and a house guest, Norma Rabinovich, 50. The Holtzbergs' cook, nicknamed Jackie, and Moshe's nanny, Sandra Samuel, were also in the house, hidden in the pantry on the first floor. All the hostages, except Moshe who escaped with his nanny, were executed.

In 2021, *The Jewish Chronicle* newspaper reported that wiretaps revealed in detail how the terrorists selected a Jewish target and other locations as part of a broader plan to strike various communities and garner maximum attention from the world's media.[31]

Today, a waterfall monument inscribed with the names of the victims of the 26/11 terror attack, including the policemen and the NSG commandos, forms a rooftop memorial at the Nariman House in Colaba. It has been rechristened as Nariman Light House.[32]

Notes

1. Sultans of the Sky

1. A club cricketer, Arthur Conan Doyle was so impressed by Nottinghamshire's fast bowler Frank Shacklock and wicketkeeper Mordecai Sherwin that he named his iconic detective character 'Sherlock', after them. 'Arthur Conan Doyle – The Game's Afoot', CricketMash, 19 January 2015, http://cricmash.com/cricket-and-literature/2015/1/19/arthur-conan-doyle-the-games-afoot.
2. 'Sussex Martlets v. Eastbourne College', Arthur Conan Doyle Encyclopedia, https://www.arthur-conan-doyle.com/images/7/74/Cricket-1911-06-17-sussex-martlets-v-eastbourne-college-p259.jpg.
3. From the time that his father had provided tuition back in Rawalpindi, Hardit Malik was to excel at sports throughout his life, particularly in tennis, cricket and golf. Stephen Barker, *Lion of the Skies: Hardit Singh Malik, the Royal Air Force and the First World War* (Kindle Edition), HarperCollins Publishers India, 2022.
4. 'When the Child of the Raj Could Find an Ever-Open Door'. Interview with Trevor Fishlock, *The Times* (London), 16 October 1982.
5. S. Barker, *Lion of the Skies: Hardit Singh Malik, the Royal Air Force and the First World War*, p. 86.
6. S. Barker, *Lion of the Skies: Hardit Singh Malik, the Royal Air Force and the First World War*.
7. 'Hardit Singh Malik', Empire, Faith and War, http://www.

empirefaithwar.com/tell-their-story/citizen-historians-in-action/soldier-stories-blog/hardit-singh-malik.

8. The Royal Flying Corps (RFC) served as the British Air Force during WWI. In 1918, it became the Royal Air Force (RAF).
9. P. Devitt and S. Sapru, 'South Asian Volunteers in the RAF', RAF Museum, 21 July 2020, https://www.rafmuseum.org.uk/blog/south-asian-volunteers-in-the-raf-part-one/.
10. P. Devitt and S. Sapru, 'South Asian Volunteers in the RAF', RAF Museum, 21 July 2020, https://www.rafmuseum.org.uk/blog/south-asian-volunteers-in-the-raf-part-one/.
11. Heike Liebau, 'Martial Races, Theory of', International Encyclopedia of the First World War, https://encyclopedia.1914-1918-online.net/article/martial_races_theory_of.
12. S. Sapru, *Laddie Goes to War: Indian Pilots in World War I*, VIJ Books India Pvt Ltd, 2021.
13. 'He Reached for the Skies: An Excerpt From Shrabani Basu's Book', *The Telegraph* (online), 8 November 2015, https://www.telegraphindia.com/7-days/he-reached-for-the-skies/cid/1313741.
14. *The Telegraph*, 'He Reached for the Skies.'
15. Ibid.
16. *The Telegraph*, 'He Reached for the Skies.'
17. S. Sapru, *Laddie Goes to War*.
18. Shrabani Basu, *For King and Another Country: Indian Soldiers on the Western Front, 1914–18*, Bloomsbury India, 2016, p. 137.
19. Md Shahnawaz Khan Chandan, 'The Mystery of the Barisal Guns', *The Daily Star*, 16 February 2016, https://www.thedailystar.net/star-weekend/strange-history/the-mystery-the-barisal-guns-1534687.
20. This conversation has been taken from S. Sapru, *Laddie Goes to War*.
21. S. Sapru, *Laddie Goes to War*.
22. 'Officer's record: Indra Lal Roy', The National Archives, UK, https://www.nationalarchives.gov.uk/pathways/firstworldwar/people/lalroy.htm

23. S. Sapru, *Skyhawks*, Writers' Workshop, Kolkata, 2007, p. 91–93.
24. Stephen Barker, *Lion of the Skies: Hardit Singh Malik, the Royal Air Force and the First World War*, HarperCollins India, 2022, p. 129 (Kindle Edition).
25. Empire, Faith and War, 'Hardit Singh Malik'.
26. S. Barker, *Lion of the Skies*.
27. Empire, Faith and War, Hardit Singh Malik.
28. This conversation has been taken from S. Sapru, *Laddie Goes to War*.
29. S. Sapru, *Skyhawks*, p. 96.
30. S. Sapru, *Skyhawks*, p. 97.
31. S. Sapru, *Laddie Goes to War*.
32. S. Sapru, *Skyhawks*, p. 98.
33. S. Sapru, *Laddie Goes to War*, pp. 112–113.
34. S. Basu, *For King and Another Country*, p. 138.
35. 'Manfred, baron von Richthofen', Encyclopedia Britannica, https://www.britannica.com/biography/Manfred-Freiherr-von-Richthofen.
36. 'Manfred, baron von Richthofen', Encyclopedia Britannica.
37. S. Basu, *For King and Another Country*, p. 138.
38. S. Sapru, *Laddie Goes to War*.
39. S. Sapru, *Skyhawks*, p. 136.
40. S. Basu, *For King and Another Country*, p. 138.
41. George Morton-Jack, 'World War One: Six Extraordinary Indian Stories', BBC News, 11 November 2018, https://www.bbc.com/news/world-asia-india-46148207.
42. 'India and UK Commemorate Fallen Soldiers in World War', Gov. UK, 9 November 2018, https://www.gov.uk/government/news/india-and-uk-commemorate-fallen-soldiers-in-world-war-1.
43. *The Telegraph*, 'He Reached for the Skies.'
44. Empire, Faith and War, 'Hardit Singh Malik'.
45. S. Basu, *For King and Another Country*, p. 184.
46. Ibid.
47. In his interview, Jagmohan Sapru said he had put the two pilots in touch. However, the phone conversation has been dramatized.

48. S. Sapru, *Skyhawks*, p. 225.
49. 'Air Marshal Subroto Mukerjee OBE', Indian Air Force, https://indianairforce.nic.in/air-m'arshal-subroto-mukerjee-obe-cas/.

2. Message in a Battle

1. Santanu Das, 'The Indian Sepoy in the First World War', British Library, 6 February 2014, https://www.bl.uk/world-war-one/articles/the-indian-sepoy-in-the-first-world-war.
2. Craig More, 'Combined Arms Warfare at Cambrai', The History Press, https://www.thehistorypress.co.uk/articles/combined-arms-warfare-at-cambrai/.
3. Lance Dafadar is equivalent to Naik in the infantry and Corporal in the British army.
4. A move order in the military refers to an official directive or instruction given to military units or personnel to relocate from one location to another.
5. *Time*, 'Books: Blood and Mud', *Time*, 13 October 1958, https://content.time.com/time/subscriber/article/0,33009,868944,00.html
6. 'Books: Blood and Mud'.
7. Léon Wolff, *In Flanders Fields: The 1917 Campaign*, Viking Press, 1958, p. 308.
8. Imperial War Museums (IWM), 'How the Battle of Cambrai Changed Fighting Tactics on the Western Front', https://shorturl.at/dnBJR.
9. Bryn Hammond, *Cambrai 1917: The Myth of the First Great Tank Battle*, Orion, 2008.
10. IWM Sound Archive, 'Kenneth Page Interview', Reference 717.
11. Graham Day, 'Battle of Cambrai 1917', British History, 18 December 2022, https://british-history.co.uk/world-war-1/cambrai-1917.
12. IWM, 'How the Battle of Cambrai Changed Fighting Tactics.'
13. Gerald Gliddon, *VCs of the First World War: Cambrai 1917*, Sutton, 2004, p. 4308.

14. B. Hammond, *Cambrai 1917*, p. 2222.
15. '1917: British Launch Surprise Tank Attack at Cambrai', history.com, https://www.history.com/this-day-in-history/british-launch-surprise-tank-attack-at-cambrai.
16. Eric Niderost, 'Tank Attack at Cambrai', Warfare History Network, June 2010, https://warfarehistorynetwork.com/article/tank-attack-at-cambrai/.
17. G. Gliddon, *VCs of the First World War*.
18. Charles Griffin, '2nd Bengal Lancers Cavalry', britishempire.co.uk, https://www.britishempire.co.uk/forces/2ndbengallancers.htm
19. C. Griffin, '2nd Bengal Lancers Cavalry'.
20. E.B. Maunsell, *Prince of Wales's Own, the Scinde Horse 1839–1922*, The Regimental Committee, 1926, p. 167; D.E. Whitworth, *History of the 2nd Lancers (Gardner's Horse) from 1809–1922*, Sifton Praed, 1924, p. 165.
21. E. Maunsell, *Prince of Wales's Own the Scinde Horse*; D. Whitworth, *History of the 2nd Lancers*.
22. C. Griffin, '2nd Bengal Lancers Cavalry'.
23. Lieutenant R. Karl Christian in *Das Heldenbuch vom Infanterie-Regiment 418*, Vereinigung ehemaliger Angehöriger IR 418, 1937, p. 125.
24. *Das Heldenbuch vom Infanterie-Regiment 418*.
25. B. Hammond, *Cambrai 1917*, p. 801.
26. C. Griffin, '2nd Bengal Lancers Cavalry'.
27. Karl Christian saw the Indian soldiers including Jot Ram and Gobind Singh approaching. He drew his pistol, waited for them to get closer, and fired. Missing the first shot and jamming his second cartridge, he was hit by a lance and fell to the ground. Injured, he survived by pretending to be dead. The horsemen then overwhelmed Christian's defences and his men. (*Das Heldenbuch vom Infanterie-Regiment 418.*)
28. C. Griffin, '2nd Bengal Lancers Cavalry'.
29. G. Gliddon, *VCs of the First World War*, p. 337. (The conversation has been imagined and dramatized for narrative purposes.)

30. B. Hammond, *Cambrai 1917*, p. 8095.
31. Defiladed is a military term used to describe a position or route that is shielded or protected from direct enemy fire or observation by natural or man-made obstacles such as terrain features, buildings or fortifications.
32. B. Hammond, *Cambrai 1917*, p. 8277.
33. Encyclopedia Britannica, 'Battle of Cambrai', https://www.britannica.com/event/Battle-of-Cambrai-1917.
34. 'British Initiative to Make Known Heroics of War Heroes', *Deccan Herald*, 17 July 2016, https://www.deccanherald.com/archives/british-initiative-make-known-heroics-2075380.
35. Shrabani Basu, *For King and Another Country: Indian Soldiers on the Western Front, 1914–18*, Bloomsbury India, 2016, p. xxi.

3. Three Lives in a War

1. Lieutenant Colonel Chanan Singh Dhillon's war diary.
2. 'The Struggle for North Africa: 1940-43', National Army Museum, https://www.nam.ac.uk/explore/struggle-north-africa-1940-43.
3. Because of large gaps in the Italian lines, British commanders launched counter attacks on isolated Italian positions, causing them to retreat. Using captured Italian tanks and artillery, the British turned these counterattacks into a major offensive, leading to the surrender of 200,000 Italians in just two weeks. (The National Archives, 'The Cabinet Papers: Egypt and Libya', https://www.nationalarchives.gov.uk/cabinetpapers/themes/egypt-libya.htm).
4. The National Archives, 'The Cabinet Papers'.
5. National Army Museum, 'The Struggle for North Africa'.
6. 1929 Geneva Convention Relative to the Treatment of Prisoners of War, 27 July 1929, sec. III, chap. 1, art. 27, p. 184, https://digital-commons.usnwc.edu/cgi/viewcontent.cgi?article=1923&context=ils.
7. 1929 Geneva Convention Relative to the Treatment of Prisoners of War, 27 July 1929, sec. III, chap. 3, art. 31, p. 184, https://digital-commons.usnwc.edu/cgi/viewcontent.cgi?article=1923&context=ils.

8. Military history fandom, 'SS Loreto (1912)', https://military-history.fandom.com/wiki/SS_Loreto_(1912)#CITEREF GreeneMassignani1994.
9. Lieutenant Colonel Chanan Singh Dhillon, 'Days of Horror', *Spectrum: The Sunday Tribune*, 16 November 2008, https://www.tribuneindia.com/2008/20081116/spectrum/main6.htm.
10. Dhillon, 'Days of Horror'.
11. Dhillon's war diary.
12. Dhillon, 'Days of Horror'.
13. Dhillon, 'Days of Horror'.
14. Dhillon's war diary.

4. The Boy Who Would Become Stak

1. 'Story of 1947 – Brutal Invasion from Pakistan in Jammu Kashmir, India Fought Back to Rescue the State', JK Now, 2019, https://www.jammukashmirnow.com/Encyc/2019/11/1/Story-of-1947-Brutal-Invasion-from-Pakistan-in-Jammu-Kashmir-India-fought-back-to-rescue-the-state.amp.html.
2. Col Bhaskar Sarkar (Retd), 'Defence of Srinagar 1947' in *Outstanding Victories of the Indian Army, 1947–1971*, Lancer Publishers, 2016. Excerpted in *Indian Defence Review*, http://www.indiandefencereview.com/interviews/defence-of-srinagar-1947/0/.
3. Sarkar, 'Defence of Srinagar 1947.'
4. Sudheendra Kulkarni, 'How and Why Gilgit-Baltistan Defied Maharaja Hari Singh and Joined Pakistan', The Wire, 23 September 2020, https://thewire.in/diplomacy/how-and-why-gilgit-baltistan-defied-maharaja-hari-singh-and-joined-pakistan
5. Kulkarni, 'How and Why Gilgit-Baltistan Defied Maharaja Hari Singh'.
6. Harbans Singh, 'Remembering the Defenders of Skardu!', Daily Excelsior.com, 13 August 2015, https://www.dailyexcelsior.com/remembering-the-defenders-of-skardu/.
7. Sirfsach.com, 'The Men Who Saved Ladakh', 24 June 2020, https://www.sirfsach.in/featured/the-men-who-saved-ladakh/24377/.

8. Lt Gen N.S. Brar, 'The Men Who Saved Ladakh', *Indian Defence Review*, 1 June 2018, http://www.indiandefencereview.com/spotlights/the-men-who-saved-ladakh/.
9. Col Ajay K. Raina and Dr Phunsog Angmo, *Lion of Ladakh: Life and Times of Colonel Chhewang Rinchen MVC, SM*, Self-published, 2023, pp. 26–27.
10. Conversations with Rinchen Angmo, daughter of Chhewang Rinchen.
11. Raina and Angmo, *Lion of Ladakh*, p. 35.
12. Raina and Angmo, *Lion of Ladakh*, p. 35.
13. Air Marshal Bharat Kumar, *An Incredible War: Indian Air Force in Kashmir War, 1947–1948*, KW Publishers Pvt Ltd, 15 October 2013, notes.
14. Brar, 'The Men Who Saved Ladakh.'
15. Raina and Angmo, *Lion of Ladakh*, p. 50.
16. Claude Arpi, 'The Unsung Heroes of the 1962 War', Rediff.com, 20 October 2022, https://www.rediff.com/news/special/claude-arpi-the-unsung-heroes-of-the-1962-war/20221020.htm.
17. Brar, 'The Men Who Saved Ladakh'.
18. Col Tej Kumar Tikoo (Retd), *1947–48 Indo-Pak War: Fall of Gilgit and Siege and Fall of Skardu*, Vivekananda International Foundation, 22 July 2013, https://www.vifindia.org/sites/default/files/1947-48-indo-pak-war-fall-of-gilgit-and-siege-and-fall-of-skardu.pdf.
19. Raina and Angmo, *Lion of Ladakh*, p. 53.
20. Brar, 'The Men Who Saved Ladakh'.
21. Brar, 'The Men Who Saved Ladakh'.
22. Maj Gen Jagatbir Singh, VSM (Retd), 'Lion of Ladakh: The Inspirational Story of Col Chhewang Rinchen', *The Sunday Guardian*, 24 December 2022 https://sundayguardianlive.com/news/lion-ladakh-inspirational-story-col-chhewang-rinchen.
23. Raina and Angmo, *Lion of Ladakh*, pp. 63–65.
24. Brar, 'The Men Who Saved Ladakh'.
25. Lieutenant Colonel Dilbag Singh Dabas (Retd), 'Maha Virs of Nubra & Kargil', *Tribune India*, 25 April 2021, https://

www.tribuneindia.com/news/features/maha-virs-of-nubra-kargil-243918.
26. Dabas, 'Maha Virs of Nubra & Kargil'.
27. Conversation with Dr Phunsog Angmo, daughter of Col Chhewang Rinchen.
28. Virendra Verma (ed.), with contributions from Saroj Kulshreshtha, *A Legend in His Own Time, Chhewang Rinchen: Memoirs*, Young India Publications, 1998, p. 101.
29. Claude Arpi, 'Have You Heard About This Indian Hero?', Rediff.com, 22 December 2011, https://m.rediff.com/news/slide-show/slide-show-1-the-hero-of-nubra/20111222.htm.
30. Raina and Angmo, *Lion of Ladakh*, p. 87.

5. Rise after the Fall of 1962

1. *Sportstar*, 'On This Day: Pakistan Ends India's Golden Run At Olympics', 9 September 2020, https://sportstar.thehindu.com/hockey/india-vs-pakistan-hockey-final-1960-rome-olympics-sports-news/article32561040.ece First published in *The Hindu* on September 9, 1960.
2. Ravindra Rai, 'Haripal Kaushik: Olympian and Decorated Solider', *Indo-Canadian Voice* [online], 1 June 2018, https://voiceonline.com/special-feature-haripal-kaushik-olympian-and-decorated-solider/.
3. *Sportstar*, 'On This Day: Pakistan Ends India's Golden Run at Olympics'.
4. Brig. I.S. Gakhal (Retd.), 'Col Haripal Kaushik, VrC, Arjuna Award: The Gentleman Soldier Sportsman', Mission Victory India, 31 January 2023, https://missionvictoryindia.com/col-haripal-kaushik-vrc-arjuna-award-the-gentleman-soldier-sportsman/.
5. John Garver, *Protracted Contest: Sino-Indian Rivalry in the Twentieth Century*, University of Washington Press, 2002, p. 28.
6. The Chinese believed the McMahon Line was a colonial legacy from the Simla Conference held from October 1913 to July 1914.
7. Arjun Subramaniam, *India's Wars: A Military History, 1947–1971*, HarperCollins, 2016, p. 211.

8. 'Sino-Indian War (1962)', Encyclopedia Britannica, https://www.britannica.com/topic/Sino-Indian-War.
9. The 4th Infantry Division was equipped and trained to fight in the plains of Punjab.
10. Conversations with Brigadier I.J. Gakhal (presently compiling a historical document on the Sikh regiment).
11. Lt Mahavir Prasad, the officer chosen, set up a post that would soon become the symbol of war.
12. Shiv Kunal Verma, *1962: The War That Wasn't – The Definitive Account of the Clash Between India and China*, Aleph Book Company, 2016, p. 191.
13. Verma, *1962: The War That Wasn't*, p. 184.
14. Mandeep Singh Bajwa, 'The Hero of Battle of IB Ridge', *The Hindustan Times*, 23 October 2012, https://www.hindustantimes.com/chandigarh/the-hero-of-battle-of-ib-ridge/story-ieEKWmL556b7TkMJGRAlkI.html.
15. Assam Rifles is a part of India's Paramilitary forces.
16. Conversations with Brig. Gakhal.
17. Conversations with Brig. Gakhal.
18. In the West, the CIA launched the Bay of Pigs operation, diverting global attention to Cuba.
19. Suchet Vir Singh, 'How Brutal Chinese Assault Across Namka Chu Drove Indian Forces Back as 1962 War Broke Out', *The Print*, 21 October 2022, https://theprint.in/defence/how-brutal-chinese-assault-across-namka-chu-drove-indian-forces-back-as-1962-war-broke-out/1176634/#google_vignette.
20. Siddharth Singh, '1962 War Debacle – The Errors Jawaharlal Nehru Made', *Mint*, 19 March 2014, https://www.livemint.com/Opinion/HCs4SMg2ojRh1T024T1KxO/1962-war-debacle-the-errors-Jawaharlal-Nehru-made.html.
21. Steven A. Hoffmann, *India and the China Crisis*, University of California Press, 1990, pp. 143–44.
22. Singh, 'How Brutal Chinese Assault Across Namka Chu Drove Indian Forces Back as 1962 War Broke Out'.
23. Singh, 'How Brutal Chinese Assault Across Namka Chu Drove Indian Forces Back as 1962 War Broke Out'.

24. Conversations with Brig. I.J. Gakhal.
25. Conversations with Brig. I.J Gakhal (presently compiling a historical document on the Sikh regiment).
26. Bajwa, 'The Hero of Battle of IB Ridge'.
27. Bajwa, 'The Hero of Battle of IB Ridge'.
28. Claude Arpi, 'Will Beijing Return the Remains of 1962 Hero?', *Deccan Chronicle*, 13 May 2021, https://www.deccanchronicle.com/opinion/columnists/120521/claude-arpi-will-beijing-return-the-remains-of-1962-hero.html.
29. Bajwa, 'The Hero of Battle of IB Ridge'.
30. The headquarters of 1st Sikh Battalion with A and B Company was located at Milakatong La, with C Company holding Pamdir and a platoon of the company under Lieutenant Baldev Randhawa, placed at Samatso. Three kilometres to the south lay Tongpen La. To the east and west of Tongpen La were rocky outcrops and sharp cliffs that added a natural defence potential.
31. Bajwa, 'The Hero of Battle of IB Ridge'.
32. https://www.gallantryawards.gov.in/awardee/2884
33. Major K.C. Praval, *Indian Army After Independence*, Lancer Publishers, 2009. Excerpted in *Indian Defence Review*, http://www.indiandefencereview.com/spotlights/1962-war-the-chinese-invasion-ii/0/.
34. Praval, *Indian Army After Independence*.
35. Praval, *Indian Army After Independence*.
36. Praval, *Indian Army After Independence*.
37. Subramaniam, *India's Wars*, p. 242.
38. Praval, *Indian Army After Independence*.
39. Praval, *Indian Army After Independence*.
40. Praval, *Indian Army After Independence*.
41. Subramaniam, *India's Wars*.
42. Conversation with Lieutenant General Kishen Khorana, veteran of the 1962 war and one of the officers on the march from Sela.
43. A vehicle column, led by the artillery unit, was followed by 4 Garhwal Rifles, 2 Sikh LI, 62 Brigade headquarters with 1 Sikh, and a few elements in the rear.

44. Conversations with Harbinder Singh, former Indian hockey international and Olympic winner.
45. Conversations with Singh.
46. Conversations with Singh.
47. Conversations with Singh.
48. Naveen Peter, 'A Stroke in Time! How Indian Hockey Team Won Gold Back from Pakistan at Tokyo 1964', Olympics, updated on 28 June 2023, https://olympics.com/en/news/indian-hockey-team-tokyo.

6. Top Guns of Boyra

1. Shamsuddoza Sajen, 'Yahya Suspends National Assembly's Inaugural Session', *The Daily Star*, 1 March 2021, https://www.thedailystar.net/backpage/news/yahya-suspends-national-assemblys-inaugural-session-2052657.
2. Sajen, 'Yahya Suspends National Assembly's Inaugural Session'.
3. FPJ Web Desk, 'What is Operation Searchlight? Know About Bloodiest Conflict in Post-WWII era', *Free Press Journal*, 24 March 2023, https://www.freepressjournal.in/world/what-is-operation-searchlight-know-about-bloodiest-conflict-in-post-wwii-era.
4. Sydney H. Schanberg, 'He Tells Full Story of Arrest and Detention', *New York Times*, 18 January 1972, https://www.nytimes.com/1972/01/18/archives/he-tells-full-story-of-arrest-and-detention-sheik-mujib-describes.html.
5. Asoke Mukerji, 'A Diplomatic Narrative of the 1971 War', *The Wire*, 18 December 2021, https://thewire.in/diplomacy/a-diplomatic-narrative-of-the-1971-war.
6. P.V.S. Jagan Mohan and Samir Chopra, *Eagles Over Bangladesh: The Indian Air Force in the 1971 Liberation War*, HarperCollins Publishers India, 2013.
7. Col Vijay Yeshvant Gidh (Retd.), 'The Battle of Garibpur', *Journal of Defence Studies*, Vol. 15, Issue 4, October–December 2021, https://www.idsa.in/system/files/jds/15_04_vijay_yeshvant_gidh.pdf.

8. Gidh, 'The Battle of Garibpur'.
9. Kabir Upmanyu, 'The Battle of Garibpur: Veterans Recall the Prelude to 1971 War', The Quint, 19 September 2021. First published 21 November 2017, https://www.thequint.com/videos/news-videos/battle-of-garibpur-a-prelude-to-india-pakistan-war-1971#read-more.
10. Lt Ajit Apte was 'On support and on call' – technical terms used to describe the type of support artillery provided the attacking units.
11. Conversations with Brig. Ajit Apte.
12. Gidh, 'The Battle of Garibpur.'
13. Conversations with Brig. Ajit Apte.
14. Pakistan tried to build another airfield in Kurmitola (now in Dhaka) to boost fighting capability.
15. Mohan and Chopra, *Eagles Over Bangladesh*, p. 82.
16. The Folland Gnat was a combat aircraft deployed in significant numbers on the eastern front. (Mohan and Chopra, *Eagles Over Bangladesh*.)
17. Mohan and Chopra, *Eagles Over Bangladesh*, p. 83.
18. Mohan and Chopra, *Eagles Over Bangladesh*, pp. 83-84.
19. Air Marshal S.Y. Savur, 'The Flight of My Life'. Unpublished manuscript.
20. Mohan and Chopra, *Eagles Over Bangladesh*, p. 84.
21. The battalion had fought the historic last-stand battle of Saragarhi against the Afghans in 1897.
22. An adjutant is an officer who assists with administrative and logistical duties within a military unit.
23. Lt Gen H.S. Panag (Retd.), 'How I Captured and Saved India's First Prisoner of War in 1971', *The Print*, 4 December 2018, https://theprint.in/opinion/how-i-captured-and-saved-indias-first-prisoner-of-war-in-1971/158246/#google_vignette.
24. Panag, 'How I Captured and Saved India's First Prisoner of War in 1971'.
25. Conversation with Group Captain Don Lazarus.
26. 'The moment the aircraft was hit, it caught fire and exploded. While the Pakistani pilot pulled the throttle to slow down and

eject, the Indian pilots following him overshot (having five times more the speed difference over the slowing Sabre after it was hit). (Conversation with Squadron Leader Sunith Soares.)

27. Air Commodore S. Sajid Haider, *Light of the Falcon: Demolishing Myths of Indo-Pak Wars: 1965 & 1971 (Story of a Fighter Pilot)*, Vanguard Books, 2009.
28. Panag, 'How I Captured and Saved India's First Prisoner of War in 1971'.
29. Panag, 'How I Captured and Saved India's First Prisoner of War in 1971'.
30. Conversations with Brig. Ajit Apte.
31. Bharat Rakshak, 'The Boyra Air Battle – 22 November 1971', 29 March 2015, Bharat-Rakshak.com, https://www.bharat-rakshak.com/IAF/history/1971war/boyra-battle/.
32. Bharat Rakshak, 'The Boyra Air Battle'.
33. Jayanta Gupta, 'How Kol Airport Facilitated Sabre Slayers in 1971', *Times of India*, 22 November 2020, https://timesofindia.indiatimes.com/city/kolkata/how-kol-airport-facilitated-sabre-slayers-in-1971/articleshow/79346589.cms.
34. Roopinder Singh, 'Flying High, Slaying Sabres', *The Tribune*, 15 April 2023 (updated), https://www.tribuneindia.com/news/musings/flying-high-slaying-sabres-497414.
35. Interview with Air Commodore Pradeep Kapoor.
36. 'How India–Pakistan War of 1971 Started, How We Won & Significance of Vijay Diwas', *Times of India*, 16 December 2022, https://timesofindia.indiatimes.com/india/how-india-pakistan-war-of-1971-started-how-we-won-significance-of-vijay-diwas/articleshow/96271054.cms.
37. History Division of the Ministry of Defence of India, 'Official History of the 1971 India Pakistan War', Bharat-Rakshak.com, 12 October 2006, http://www.bharat-rakshak.com/archives/OfficialHistory/1971War/1971Chapter10.pdf.
38. Man Aman Singh Chhina, 'Explained: How the Indo-Pak War of 1971 Began 50 Years Ago on This Day', *Indian Express*, 4 December 2021, https://indianexpress.com/article/explained/india-pakistan-1971-war-events-explained-7653653/.

39. Chhina, *Indian Express*.
40. Online memorial: Flight Lieutenant Pradeep Vinayak Apte VrC, Honourpoint, https://www.honourpoint.in/profile/flight-lieutenant-pradeep-vinayak-apte-vrc/.
41. Air Commodore Nitin Sathe, '1971 War: Solitary Confinement In Pakistan', Rediff.com, 16 December 2022, https://www.rediff.com/news/special/-1971-war-flying-officer-jawahar-lal-bhargava-solitary-confinement-in-pakistan/20221216.htm.
42. Conversations with Brig. Ajit Apte.
43. *Times of India*, 'How India-Pakistan War of 1971 Started, How We Won & Significance of Vijay Diwas'.
44. *New York Times*, 'India to Release 90,000 Pakistanis in Peace Accord', 29 August 1973, https://www.nytimes.com/1973/08/29/archives/india-to-release-90000-pakistanis-in-peace-accord-hardsought.html.
45. Lieutenant Colonel M.K. Guptaray (Retd.), 'Missing 54 Indian POWs of the 1971 War', Mission Victory India, 15 March 2022, https://missionvictoryindia.com/missing-54-indian-pows-of-the-1971-war/.
46. Don Lazarus served as a flight instructor for instrument ratings and was a pioneering member of India's first electronic countermeasure (ECM) squadron, a specialized unit dedicated to disrupting enemy radar and communication systems.
47. Bharat Rakshak, 'The Boyra Air Battle'.
48. Conversations with Group Captain Don Lazarus.
49. Air Commodore Kaiser Tufail, 'Himalayan Showdown: Kargil Crisis 10 Years On', *AirForces Monthly* (UK), Key Publishing Ltd., June 2009.

7. A Bloodless Pact to Victory

1. 'All You Need to Know About Kargil War', *Economic Times*, 26 July 2017, https://economictimes.indiatimes.com/news/defence/all-you-need-to-know-about-kargil-war/casualties/slideshow/59772212.cms.

2. This story is based on the account of Major Ajit Singh (Retd)
3. Marpo La ridge originates from Pakistan Occupied Kashmir and comprises several peaks including Pt 5240. The ridge overlooks the Drass sector in India.
4. Aziz, the Pakistani commander, is believed to be a colonel.
5. The India-Pakistan Simla agreement, 2 July 1972, https://www.mea.gov.in/Portal/LegalTreatiesDoc/PA72B1578.pdf.
6. Devdas was a famous literary character from Sarat Chandra Chattopadhyay's novel by the same name, known for his pining in love.
7. During a battle, an infantry officer at the front observes the enemy positions, providing critical information to artillery guns placed in the rear regarding the enemy's distance and direction. Subsequently, the artillery can bombard and neutralize the enemy defences, facilitating the infantry in launching their attack.
8. Ashraf Wani, 'Bofors Power Proved in Kargil War', *India Today*, 25 July 2016, https://www.indiatoday.in/india/story/kargil-winning-the-battle-with-bofors-331307-2016-07-25.

8. Warrior's Code of Courage

1. 'Chronology of ULFA since 1979', *India Today*, 4 December 2009, https://www.indiatoday.in/india/story/chronology-of-ulfa-since-1979-62301-2009-12-03.
2. Farzand Ahmed 'Operation Rhino: Indian Army Cracks Down ULFA Activities, Arrest Top Leaders', *India Today*, 31 January 1992, https://www.indiatoday.in/magazine/indiascope/story/19920131-operation-rhino-indian-army-cracks-down-ulfa-activities-arrest-top-leaders-765740-2013-06-29.
3. The Eastern Command includes the territories of Sikkim, Arunachal Pradesh, Nagaland, Manipur, Mizoram, Tripura, Meghalaya, Assam and Bengal. (Indian Army, 'Eastern Command', https://indianarmy.nic.in/command/command/eastern-command-commands-site-main).

4. Avirook Sen, 'It's a Life of Carrom and Comfort for 51 ULFA Militants Who Surrendered Recently,' *India Today*, 10 August 1998, https://www.indiatoday.in/magazine/states/story/19980810-its-a-life-of-carrom-and-comfort-for-51-ulfa-militants-who-surrendered-recently-826854-1998-08-09.
5. Indian Strategic Knowledge Online, 'OP Vijay', https://indianstrategicknowledgeonline.com/web/Kargil
6. The term 'personal weapon' in the army denotes the designated weapon an officer is assigned to carry.
7. Lieutenant Colonel Manoj K. Channan (Retd.), 'Life of a Soldier and Veterans' Code of Conduct', *Financial Express*, 18 November 2019, https://www.financialexpress.com/business/defence/life-of-a-soldier-and-veterans-code-of-conduct/1768270/.

10. Nariman House, 26/11: A War Comes Home

1. Sandeep Unnithan, *Black Tornado: The Three Sieges of Mumbai 26/11*, HarperCollins Publishers India, 2014.
2. Unnithan, *Black Tornado*.
3. 'Who Was Hemant Karkare, the Lead Investigator in the 2008 Malegaon Blast?', *Indian Express*, Updated 19 April 2019, https://indianexpress.com/article/who-is/who-was-hemant-karkare-the-lead-investigator-in-the-2008-malegaon-blast/.
4. 'Timeline: Mumbai Assault', Al Jazeera, 26 November 2009, https://www.aljazeera.com/news/2009/11/26/timeline-mumbai-assault.
5. Unnithan, *Black Tornado*.
6. Unnithan, *Black Tornado*.
7. The concept of the Balidaan badge was adopted by the NSG during a period when the Para Special Forces (SF) made up 70 per cent of its composition. While the SF uses a dagger as their Balidaan insignia, the NSG features a trident with wings as their Balidaan badge emblem. The NSG's specialized training focuses on mastering various weapons for specific roles tailored to its needs, and this training differs from that of the SF. This distinction in

training is why the Balidaan badge is a separate and unique symbol for the NSG.
8. Unnithan, *Black Tornado*.
9. Conversations with Lieutenant Colonel Sundeep Sen.
10. Conversations with Namrata Sen.
11. Ashish Khetan, *26/11 Mumbai Attacked*, edited by Harinder Baweja, Lotus Collection (Roli Books), 2009, p. 71.
12. Jeremy Kahn, 'Jews of Mumbai, a Tiny and Eclectic Group, Suddenly Reconsider Their Serene Existence', *New York Times*, 2 December 2008, https://www.nytimes.com/2008/12/03/world/asia/03jews.html.
13. Unnithan, *Black Tornado*.
14. 'Explosive Audio Tapes Expose 26/11 Terror Plot', *India Today* TV Tapes, YouTube, 8 February 2016, https://m.youtube.com/watch?v=D5jW5C-fqrA&feature=youtu.be.
15. *India Today* TV Tapes.
16. Unnithan, *Black Tornado*, p. 189.
17. Unnithan, *Black Tornado*, p. 193.
18. India Today TV Tapes, https://m.youtube.com/watch?v=D5jW5C-fqrA&feature=youtu.be.
19. Unnithan, *Black Tornado*, p. 194.
20. Conversation with sources.
21. Unnithan, *Black Tornado*.
22. The use of phones was not unusual. Without reliable intelligence, coordinating resources and operational planning was being done outside the building.
23. Khetan, *26/11 Mumbai Attacked*, p. 78.
24. Conversations with sources close to NSG.
25. Unnithan, *Black Tornado*, p. 196.
26. Unnithan, *Black Tornado*.
27. Unnithan, *Black Tornado*.
28. Krishna Kumar, 'Did Patil delay the NSG?', *India Today*, 4 December 2008, https://www.indiatoday.in/latest-headlines/story/did-patil-delay-the-nsg-34388-2008-12-02.
29. Baweja (Ed.), *26/11 Mumbai Attacked*.

30. Unnithan, *Black Tornado*.
31. Ben Felsenburg, 'Wiretap Recordings of Mumbai Attackers Reveal How Chabad House Was Chosen as Jewish Target', *The Jewish Chronicles*, 17 December 2021, https://www.thejc.com/news/world/wiretap-recordings-of-mumbai-attackers-reveal-how-chabad-house-was-chosen-as-jewish-target-1xTtzmFdwP99DEagcQBjX3.
32. 'Nariman Light House: From Darkness to Beacon of Light', *The Hindu*, 25 November 2018, https://www.thehindu.com/news/cities/mumbai/nariman-light-house-from-darkness-to-beacon-of-light/article62022569.ece.

Select Bibliography

1. Sultans of the Sky

1. Somnath Sapru, *Laddie Goes to War: Indian Pilots in World War I*, VIJ Books India Pvt Ltd, 2021.
2. Stephen Barker, *Lion of the Skies: Hardit Singh Malik, the Royal Air Force and the First World War*, HarperCollins India, 2022.
3. Somnath Sapru, *Skyhawks*, Writer's Workshop – Songbird Books, 2006.
4. Shrabani Basu, *For King and Another Country: Indian Soldiers on the Western Front, 1914–18*, Bloomsbury India, 2016.
5. Conversations with Somnath Sapru.
6. Wisden Cricket website.
7. The National Archives, Surrey, UK.
8. Balliol College Archives, Oxford University.
9. Hardit Singh Malik, *A Little Work, A Little Play: The Autobiography of H.S. Malik*, Bookwise (India), 2009.

2. Message in a Battle

1. Léon Wolff, *In Flanders Fields: The 1917 Campaign*, Viking Press, 1958.
2. Bryn Hammond, *Cambrai 1917: The Myth of the First Great Tank Battle*, Orion, 2008.
3. Gerald Gliddon, *VCs of the First World War: Cambrai 1917*, Sutton, 2004.

4. D.E. Whitworth, *History of the 2nd Lancers (Gardner's Horse) from 1809–1922*, Sifton Praed, 1924.
5. Inputs from Colonel Rajinder Singh.
6. Andrew Rawson, *The Cambrai Campaign 1917*, Pen and Sword Books Ltd, South Yorkshire, 2017.
7. First World War: War Diary (France, Belgium and Germany), The Naval and Military Press Ltd (National Archives), 1914–1922.

3. Three Lives in a War

1. War Diary maintained by Lieutenant Colonel Chanan Singh Dhillon.
2. Conversations with Gurbinder Singh Dhillon.

4. The Boy Who Would Become Stak

1. Virendra Verma (ed.), with contributions from Saroj Kulshreshtha, *A Legend in His Own Time, Chhewang Rinchen: Memoirs*, Young India Publications, 1998.
2. Col Ajay K. Raina and Dr Phunsog Angmo, *Lion of Ladakh: Life and Times of Colonel Chhewang Rinchen MVC, SM*, Sabre and Quill Publishers, New Delhi, 2023.
3. Conversations with Dr Phunsog Angmo.

5. Rise after the Fall of 1962

1. Conversations with Brig. I.S. Gakhal (Retd.) and notes from his unpublished manuscript on the period.
2. Major Sitaram Johri, *Chinese Invasion of NEFA*, Himalaya Publications, 1968.
3. Shiv Kunal Verma, *1962: The War that Wasn't*, Aleph Books, 2016.
4. Conversations with Lieutenant General Kishen Khorana.
5. Conversations with Major General Bhullar.
6. Conversations with Colonel Avinash Rayarikar.

6. Top Guns of Boyra

1. P.V.S. Jagan Mohan and Samir Chopra, *Eagles Over Bangladesh: The Indian Air Force in the 1971 Liberation War*, HarperCollins Publishers India, 2013.
2. Conversations with Brigadier Ajit Apte, Squadron Leader Sunith Soares, Wing Commander Don Lazarus and Squadron Leader Pradeep Kapoor.
3. Lieutenant General H.S. Panag (Retd.), 'How I Captured and Saved India's First Prisoner of War in 1971', *The Print*.

7. A Bloodless Pact to Victory

1. This chapter is based on correspondence and notes of Major Ajit Singh (Retd). The information shared by these sources has been used with their permission and with the understanding that their identity will remain confidential.

8. Warrior's Code of Courage

1. Conversations with Brigadier Saurabh Singh Shekhawat.

9. The Militant and the Major

1. This chapter is based on confidential conversations with sources who have chosen not to be named. The information shared by these sources has been used with their permission and with the understanding that their identity will remain confidential.

10. Nariman House, 26/11: A War Comes Home

1. Sandeep Unnithan, *Black Tornado: The Three Sieges of Mumbai 26/11*, HarperCollins Publishers India, 2014.
2. Harinder Baweja (ed.), *26/11 Mumbai Attacked*, Lotus Collection (Roli Books), 2009.

3. 'Explosive Audio Tapes Expose 26/11 Terror Plot', *India Today* TV Tapes, https://m.youtube.com/watch?v=D5jW5C-fqrA&feature=youtu.be.
4. Conversations with Colonel Sundeep Sen and Namrata Sen.

Acknowledgements

One evening, as I sat down to write a story for this book, set in the Kashmir valley, I received a call from a friend posted on the India-Pakistan Line of Control. He sounded excited at the prospect of reading unique accounts of soldiers on the front such as him. On another occasion, when I was in Kalimpong, invited by the erudite Major General Khaled Zaki to speak about my book *Watershed 1967*, the general reminisced warmly about the use of 'aankhon dekha haal' (eye witness account) in the lesser known stories from that book. Separately, over a conversation, a noted economist friend told me he believed that meaningful war stories were ironically anti-war. I realized that a few stories in this book *Camouflaged* delve into the desolation of war. Besides, several such conversations endorsed the idea of different perspectives of war that this book has sought to cover. A huge thanks to all my friends.

At the beginning, the idea of stories that humanizes war narratives appealed to Kanishka Gupta, my literary agent and Parth Mehrotra, Editor in Chief, Juggernaut Books. I am obliged to them for their trust at an early stage. The work on the book, which began during Covid-19, involved research, travel, interviews and chance conversations. To understand the context, I drew upon the works of writers such as Stephen Barker, Shrabani Basu, George Morton-Jack, Arjun Subramaniam, Samir Chopra, Jagmohan Sapru, Santanu Das among others, and sources from the Imperial War Museum Archives, London and Nehru Memorial Museum and Library.

I owe my appreciation to Brigadier Arif Shaikh and Colonel Rajendra Singh Rathore, grandson of Lance Daffadar Gobind Singh for the incredible story of Cambrai. Jagmohan Sapru patiently answered my questions on India's groundbreaking fighter pilots which helped me hone the first story in the book. I was pained to hear about his passing during the research of the book. I owe much gratitude to him.

It is not easy to find an unpublished story from the World Wars. Air Vice Marshal Arjun Subramaniam suggested a story he believed had potential. Thus, Lieutenant Colonel Chanan Singh Dhillon's scintillating story, using his notes in his prison diary from WWII, took shape. Thanks to Chanan's son Gurbinder Dhillon and his sisters Satwant, Satvinder, Baljinder, Taran and Gurmeet who translated his notes from Punjabi to English, which they generously shared with me. The story wouldn't have been possible without Gurbinder's unfailing ability to gather information and corroborate diary notes. I can't thank him enough for his tireless efforts.

An unforgettable part of writing involved travel to places and meeting fascinating people. It was mesmeric listening to stories from Dr Phunsog Angmo, Brigadier Saurabh Shekhawat in Ladakh and Ajit Singh in Meghalaya surrounded by a magnificent landscape. It was a treat to listen to Ajit, a natural storyteller, while Saurabh's accounts brought alive the incident.

Listening to an emotional Lieutenant General Kishen

Acknowledgements

Khorana at his home one evening in Chandigarh, while he narrated his poignant experience from Sela in the 1962 war, will remain etched in my memory. I am grateful to India's former hockey star Harbinder Singh, the tireless Colonel Avinash Rayarikar, Brigadier Chimni and Veronica Kaushik for their detailed inputs on the story of Haripal Kaushik. I would especially like to thank Brig Injo Gakhal for sharing his detailed understanding of the 1962 India-China conflict. Past works of eminent historians including Shiv Kunal Verma and Claude Arpi were instrumental in setting the context.

The story of air force pilots in 1971 involving a dogfight and its unique aftermath couldn't have struck the right notes without the accounts of Squadron Leader Sunith Soares in Mumbai and Brigadier Ajit Apte in Pune. Discussions with Wing Commander Don Lazarus and Squadron Leader Pradeep Kapoor alongside Lieutenant General HS Panag's article helped gather a charming story.

I discovered the book's final story during a chance conversation at a regimental gathering in Delhi, which led me to the foothills of Kumaon. Lieutenant Colonel Sundeep Sen and his wife Namrata Sen shared a spellbinding account of an unknown story in the 26/11 terror attacks. Thanks to journalists, historians and authors whose inputs have been helpful but whose names may have inadvertently got missed out. I am obliged towards Sandeep Unnithan's excellent work on the events of 26/11 that helped me understand the context. There are some sources in the book, who would prefer to stay unnamed. I am equally grateful to them for reposing faith in me.

My literary agent Kanishka Gupta is my go-to person for all questions. Grateful to him and Amish Mulmi who provided valuable insights on the manuscript. Thanks to Nina Xanthe James in Sydney, Donata Afonso in London, Sunita Wazir, Colonel Akhilesh Tomar, Arayna DasGupta and Mangesh for uncomplainingly reading the stories and offering precious suggestions. I had the privilege to seek out Air Vice Marshal Arjun Subramaniam for his sagacious advice on the book. Major Manish Mall has been kind-hearted with sharing his readings on military history. My thanks to ADGPI branch (Army HQ) for reading the draft of one of the stories. Grateful to my friend and journalist Sriram Karri, whose uncanny knack of finding contemporary relevance to historical subjects is an opinion I trust. Grateful for the treasured support of Chiki Sarkar and her team at Juggernaut Books. The book wouldn't have been possible without the expertise of Parth and his dedicated team – Sonal Nerurkar, Sulagna Nandi and Yashika in shaping the narrative into an engaging form. Thankful to Arani Sinha and her team for a scrupulous approach to the production of the book. And thanks to Hassan for the maps.

I cannot forget my talented and extremely efficient research assistants: Mehr helped with research and choice of sources for the stories and Zoya Gill's meticulous work helped identify inconsistencies and mould the draft. Thanks to the resourceful Sumitra Ray who I could count for help on information and sources at several stages. The support of my friends, batchmates in uniform and family after my first book *Watershed 1967* provided the momentum for this one. My regiment – 11 Gorkha Rifles – will always be my home of dependable support; Lieutenant General Rakesh Sharma and Lieutenant General KVS Lalotra have been stellar sources of inspiration. A special personal gratitude to General Bipin Rawat, India's first Chief of Defence Staff (CDS),

Acknowledgements

who encouraged me to write more, but passed away in an unfortunate accident in December 2021.

Thanks to friends Rahul Sharma, Sue Aguilar, Stephen Weinberg, Tushar Nemade, especially Maroof Raza for the faith and support. My teachers and classmates from my school, Daly College, including Mr. Ahmed Ansari, Shamit Dave, Sangram Singh, Aparna Bidasaria and others are proof that an unwavering vote of confidence helps a long journey. I am grateful to my batchmates from the Indian Military Academy – 94th course, especially Ajit, Brig Samar Pundir, Brig Shekhawat, Sudesh Dhanda, Brig Arif and everyone else who shared every story they knew of. A deep, personal gratitude for a batchmate and friend, the wonderful Colonel Khushwant Singh Chaddha who was keen to read the book, but was sadly taken away from us during Covid-19.

Writing and travel, alongside my corporate career, took a substantial part of my personal time and I am grateful to my family for their patience. Thanks to my children Arayna and Siddharth, my mother Shyamla, my mother-in-law Lathika and my brother Protip for their belief in me. I am indebted forever to my father, Bipradas, who isn't around to read my books but is the biggest inspiration. This journey couldn't have been possible without his blessings.

My wife Nisha bore the brunt of my eccentric story ideas and yet decided to soldier on. It is apt that I dedicate this book to her and the unsung spouses of those who serve in the line of duty and shape the many 'Camouflaged' stories from battlefields.

A Note on the Author

Probal DasGupta is an author and columnist.

He has previously served as an infantry officer in the Indian Army.

Probal is the author of *Watershed 1967: India's Forgotten Victory Over China* (Juggernaut Books). This is his second book.

An alumnus of Daly College and Columbia University, New York, he has been a Braun-Meyers Fellow, JN Tata Scholar and RD Sethna Scholar. He has led global consultancies in South Asia, advising firms on reputation and business risks.

http://www.probaldasgupta.com